AIRCRAFT HYDRAULICS

Prepared by

NAVAL AIR TECHNICAL TRAINING COMMAND

for publication by the

BUREAU OF NAVAL PERSONNEL

NAVAL TRAINING COURSES

NAVPERS 10332–A

UNITED STATES
GOVERNMENT PRINTING OFFICE
WASHINGTON: 1951

For talc by the Superintendent of Document., U.S. Government Printing Office Washington 25, D. C. - Prict 75 conts

PREFACE

This book is written for enlisted men of Naval Aviation. It is one of a series designed to furnish the special-'ized information they will need to perform their highly technical duties in aviation.

The object of this book is to explain the operation of aircraft hydraulic equipment so that the Aviation Structural Mechanic, who must work with this equipment, may understand the operation of the essential units, and having this understanding, may more efficiently carry out service activities. Qualifications for advancement in ratings for the AM are included at the back of this book in appendix II.

Likewise this book will prove valuable to the Aviation Electrician's Mates—the AE and AEI—who are required by their qualification .211 to know the "principles of hydraulics applicable to aircraft instruments."

Hydraulics is a study of fluid in motion. Hydraulic systems in aircraft are used to operate a great number of vital components of the plane, and the text of this book is devoted to the study of hydraulics as used in such gear.

This study begins with an introduction to and history of aircraft hydraulics, and proceeds with a discussion of the basic principles of hydraulically actuated mechanisms. The AM is then directed to the elementary phase of hydraulic fluids and fittings. Ensuing sections include discussions of line maintenance and trouble shooting, and the book concludes with a chapter on hydraulic automatic pilots.

These subjects are presented in a simplified, copiously illustrated manner, giving thoroughly the information which the mechanic must have but eliminating that which will be of little practical value. The illustrations contained are schematic and are intended to show the operation rather than engineering drawings of the actual units.

As one of the Aviation Series of Navy Training Courses, this book represents the joint endeavor of the Naval Air Technical Training Command and the Training Division of the Bureau of Naval Personnel.

READING LIST

NAVY TRAINING COURSES
Airplane Structures, NavPers 10331
Aircraft Materials, NavPers 10330
Aircraft Metal Work, NavPers 10323
Aircraft Welding, NavPers 10322-A
Introduction to Airplanes, NavPers 10303
Mathematics, Vol. 1, NavPers 10069
Mathematics, Vol. 2, NavPers 10070
Blueprint Reading and Layout Work, NavPers 10305
Hand Tools, NavPers 10306

USAFI TEXTS
United States Armed Forces Institute (usafi) courses for additional reading and study are available through your Information and Education Officer.* A partial list of those courses applicable to your rate follows:

Number Title
J 204 Airplane Maintenance I

J 205 Airplane Maintenance II

J 362 Arc Welding

J 363 Gas Welding

• "Members of the United States Armed I V c -vc;*erve Components, when on active duty, are eligible to enroll j8A7! courses, services, and

materials if the orders calling them to v.tive duty specify a period of 120 days or more or if they have been on active duty for a period of 120 days or more, regardless of the time specified in the active duty orders."

CONTENTS

CONTENTS—Continued

VIM
AIRCRAFT HYDRAULICS

CHAPTER 1
INTRODUCTION TO HYDRAULICS DEFINITION

When you wanted to learn what made the kitchen clock tick, you simply took it apart. To determine the meaning of hydraulics, an application of that same principle will furnish you with a clearer understanding of the term.

First cut the word hydraulics in two and examine the severed sections. Apparently, the Greeks had two words for it. One end of the word came from the Greek term hydro, meaning water; the other is derived from AULIS, denoting pipe, or tube.

You know that water is a liquid, and that a tube is a container. By putting the two Greek words together again, you put liquid into a container and thereby create an English term meaning confined liquid —the literal translation of our word hydraulics.

WORKING MODEL

To gain a clear and concise understanding of the term hydraulics, let us use an ordinary dime-store water pistol as a working model. When the barrel of the pistol is filled with water, you have an example of a confined liquid.

You have liquid in a tube. When you pull the trigger, you have a practical demonstration of hydraulics in its simplest form. This action has an effect on the confined liquid. You put the fluid in motion. You exerted pressure upon the "hydro" in the "aulis," and in so doing, you create the principle of hydraulics.

But our experiment with the water pistol has accomplished little beyond splattering something or someone with water. Suppose, however, that the force expended in projecting water from the barrel of the pistol had been put to practical use; suppose that force had been turned to mechanical advantage. If the force had been used to actuate certain hydraulic units— such as automatic pilots, landing gear, or bomb-bay doors—we would have seen hydraulics functioning as the highly developed mechanical art it has become.

Since such consideration is our purpose, we may as well summarize here and now by defining hydraulics as

A SCIENCE DEALING WITH THE TRANSMISSION OF ENERGY THROUGH FLUIDS UNDER PRESSURE.

BACKGROUND OF HYDRAULIC PRINCIPLES

We have seen that the Greeks gave us our word, and it was also a Greek who first formulated the principle of HYDROSTATICS which is the science of liquids at rest. That was the work of Archimedes, who lived about 250 B. C. Little additional progress was made in the science until some 300 years ago, when, in 1612, an Italian scientist, Galileo, wrote an article in which he mentioned the Hydrostatic Paradox.

Since a paradox is anything which apparently contradicts itself, Galileo's brainchild seems to be a fact which "just ain't so." In the form of a riddle, he attempted the explanation of one of the chief properties of a liquid. Let's examine his idea as illustrated in figure 1.

The two containers, A and B, are of equal size and height. But A has a base of one square inch, while B has

Figure 1-Hydrostatic Paradox.

a base of three square inches. The containers hold equal amounts of liquid, and each container, when filled, weighs 3 pounds. If the height of A is such that the weight of the liquid causes a unit pressure of 2 pounds per square inch on the base, what will be the total force pushing against the bottom of B?

The answer is 6 pounds, and therein lies the paradox. For that amount of force is considerably greater than the total weight of the liquid. Although the base of B is under the same pressure as the base of A, there are three 1-inch areas in B. There is a total force, therefore, three times as great as that exerted upon A, making 6 pounds. This is known as the Hydrostatic Paradox.

EVERYDAY HYDRAULICS

Many instances in which the laws of hydraulics are applied to practical advantage exist in everyday life. In figure 2, such an instance is illustrated. It demonstrates the manner in which mechanics often remove a "frozen" bushing. Pressure from the hammer blow, transmitted to the fluid, is multiplied and exerted upward against the bushing. It is quite possible that the mechanic employing this method fails to realize that he has put principles of hydraulics to work.

The word hydraulics is as familiar to us today as "singletree" and "hame-strap" were to our grandparents. Most of us have driven automobiles equipped with "hy-dromatic" transmissions; stopped them with hydraulic

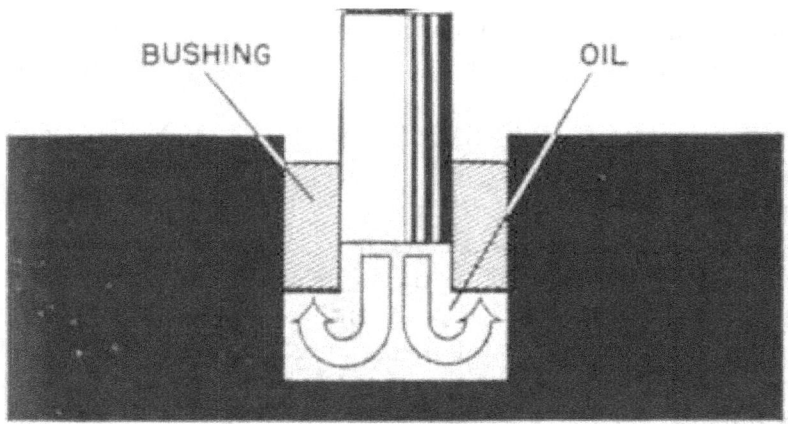

Figure 2.—Removing a "frozen" bushing.

brakes; changed tires with the help of hydraulic jacks, and had our cars greased on hydraulic lifts. A prominent automobile manufacturer recently predicted that automobiles completely equipped hydraulically will soon be as common as Sunday drivers. All this indicates that hydraulics is destined to have even greater application in industry.

HYDRAULICS IN INDUSTRY

The application of hydraulics to the aircraft and automotive industries is, however, of comparatively recent origin. Years before planes dropped from the skies on hydraulically equipped landing gear, or before an automobile stopped on a dime when hydraulic brakes were applied, the uses of hydraulics were evident to far-sighted leaders of industry.

Long before the world became air-minded, industry needed tremendous power to shape metals for use in the manufacture of machinery and equipment. Mechanical arrangements then in use were heavy, bulky, and inefficient because of the high toll exacted by friction, gearing, and backlash. With the advent of hydraulic power, however, undreamed-of pressures became not only possible but commonplace through the use of simple, easily operated equipment. Forces of 420 tons (840,000 pounds) on a flanging press, and 13,000 tons (2,600,000 pounds) on a forging press, have been hydraulically obtained.

Today, coal hoists, cranes, riveters, rams, and numerous types of remote control units are actuated by hydraulic transmissions of power.

HYDRAULICS IN THE NAVY

Before narrowing our discussion to its application to aircraft, let us consider several examples of the manner in which the Navy has harnessed the horses of hydraulics.

The guns of a battlewagon are the reasons for the big ships' existence, and hydraulics plays a major role in their deadly efficiency. Hydraulics supplies the force directing the potent voices of our fighting ships, and the terrific recoil of those guns is absorbed on hydraulic cushions.

And deep in the bowels of our submarines, hydraulics operates the fat, shiny cylinders responding to the terse command: "Up periscope!"

These are but a paltry few of the many duties the Navy assigns to hydraulics. Coming nearer home, let's glance briefly at the Navy's air arm. What raises aircraft to the flight decks of our carriers? What absorbs the shock of a landing plane hooking its arresting gear to a carrier deck pendant? The answer is hydraulics, and the listing of such examples could be continued ad infinitum.

From all that has been said, it is but logical to draw the conclusion that enormous emphasis has been, is being, and will be placed upon the importance of hydraulics. But by no stretch of the imagination is it to be inferred that the uses of other modes of power have been intentionally minimized. Entirely to the contrary other means of transmitting energy—such as mechanical, electrical, and pneumatic—are not only valiant allies of hydraulics, but in many instances are harnessed together with hydraulics to form an efficient team. An example of such coordination may be observed in the use of warning systems, remote-control selector valves, and other electro-hydraulic and hydro-pneumatic units. An excellent illustration is the "gunfire interruptor" used on modern hydraulic turrets.

HYDRAULICS IN THE AIR

Hydraulics is in its element in aircraft. Let us consider the uses to which it is put, and the advantages it offers over other modes of power.

What are the functions performed by hydraulics aboard a modern airplane? Basically, a hydraulic system is designed to operate certain mechanisms for the pilot and crew. Specifically, hydraulically actuated mechanisms do just about everything except write letters.

Hydraulics folds the wings, applies the brakes, lowers the wing flaps, raises the landing gear, opens the cowl flaps, releases the arresting hook, changes the propeller pitch, opens the bomb-bay doors, releases the bombs, charges the guns, rotates the gun turrets, helps the pilot overcome high control forces on modern high-speed air- • planes, and actually flies the plane by operating the mechanical pilot.

So it is evident that hydraulics is a "must" in aviation, and it stands to reason that if it can be made to perform such a wide variety of jobs so very well, it must certainly be good. Now, our problem is to determine what makes it good.

In analyzing the advantages of hydraulics, the best method of approach will be to draw comparisons between hydraulics and other methods of energy transmission. So let's put the two other methods—mechanical and electrical—side by side with hydraulics and see where advantages or disadvantages for each exist.

HYDRAULICS VS MECHANICAL

How does a hydraulic system compare with one operated mechanically?

First of all, consider the element of efficiency. The principal charge against any mechanical unit from the outset is its inefficiency. Mechanical systems deliver, roughly, between 60 and 70 percent of the energy that is put into them, this figure varying according to the accuracy with which parts are made, tolerances, modes of lubrication, and the like. On the other hand, the efficiency of a hydraulic system, discounting linkages, is almost 100 percent. The slight loss (a fraction of 1 percent) is due to internal friction in the liquid and in the devices.

Then there is the matter of installation. In any mechanical system, a series of more or less straight lines are essential. This would prove a serious handicap in a device with intricate bracing such as is contained in an airplane. Space in aircraft is extremely limited, and few straight lines can be maintained for any appreciable distance. Hydraulic lines, in comparison, may be bent almost at will around any obstruction.

Weight is another important consideration. No slide-rule computation is required to prove that weight is always the prime factor in aircraft. The average mechanical installation is bulky and, consequently, very heavy. A strand of flexible or extra-flexible cable 15 or 20 feet in length has considerable weight when contrasted with small, thin-walled aluminum tubing.

In comparing mechanical with hydraulic systems, contrasts in wear and lubrication also demand close scrutiny. Mechanical systems require frequent lubrication, and wear due to friction necessitates constant replacement of parts. In the hydraulic system, however, moving parts are at a minimum, and lubrication is continuously supplied by the liquid itself.

HYDRAULIC VS ELECTRICAL

There can be little argument against the fact that electricity was an extremely valuable means of transmitting energy. But once again you are reminded that our chief concern is with energy transmission peculiarly adaptable to aircraft.

As always, weight is the first item for our consideration. And once again, weight—or the lack of it—is heavily on the side of hydraulics.

On a unit-for-unit basis, the size and weight of an electric motor required to perform a certain task will be greater than that of an actuating cylinder necessary for the same purpose. With the development of 3,000-pound systems, hydraulics will perform an even greater feat in this respect.

Another factor high on the list of importance is inertia. You will recall from your high school physics that inertia is the tendency of a body at rest to remain at rest—or a body in motion to continue in motion—at a constant velocity unless acted upon by an outside force. The matter of bodies in motion commands our attention at the moment.

We know that the power of an electric motor is derived from the high speed at which its armature rotates. We

know also that it is next to impossible to stop instantly this rapidly-spinning mass, because of inertia. This proves a great disadvantage when a large moving part— such as a wing which is being folded or a landing gear which is being lowered—is brought to a sudden stop.

SUMMARY

Thus we see that hydraulic power systems are efficient, light in weight, reliable, compact, easy to install and maintain, safe in operation, adaptable to practically all types of actuation, rapid and smooth in action, not prone to overtravel, capable of instant starting and stopping, and, finally, comparatively inexpensive.

All in all, hydraulics is so very well adapted to aircraft use that most of the large manufacturers designing and producing aircraft for the Navy have accepted hydraulics as the standard mode of energy transmission.

QUIZ

1. By what name is "the science of liquids at rest" known?

2. What term is defined as "a science dealing with the transmission of energy through fluids under pressure"?

3. Name four means of transmitting energy.

4. Name 13 functions performed by hydraulics aboard a modern airplane.

5. Name five major elements of comparison which make hydraulic systems superior to mechanical systems in modern airplanes.

6. What two factors make hydraulic systems superior to electrical systems for installations on airplanes where the job involves energy transmission?

r

CHAPTER 2
HYDRAULIC PRINCIPLES
FLUIDS AND LIQUIDS
When we speak of principles from a scientific viewpoint, we are referring to the natural

laws governing the actions of a particular substance under specific and fixed conditions. In the case under consideration, that substance is a liquid.

Liquids have very definite reactions, generally expanding when heated and contracting when cooled. These set reactions, whether determined visibly or otherwise, are called hydraulic principles.

According to its definitions, a fluid is anything with the ability to flow, and since a fluid may be either liquid or gas, we will consider both types.

A liquid is a fluid whose particles have freedom of movement among themselves but lack a tendency to separate. A liquid also may be defined as a fluid of definite volume that will take the shape of its container.

By definite volume, we mean that any quantity, such as a quart, will remain a quart whether put into a quart container, a gallon container, or a 5-gallon container. In other words, a liquid has definite volume but no definite shape.

In an aircraft hydraulic system, a liquid flowing through tubing is used to transmit force from a pump to the point where the force is to be applied. Although fluid and liquid do not mean the same thing, it is customary among aviation personnel to use the two words interchangeably.

FLUIDS IN HYDRAULICS

As our discussion progresses, it will become second nature to associate "hydraulic fluids" with every phase of hydraulics. This, then, is the time to emphasize that the fluid used in hydraulic systems is always a liquid, except in extremely rare cases when compressed air or nitrogen, or even carbon dioxide gas, may be used for operation of some important unit such as the landing gear. It will be recalled that the word "hydraulic" refers to liquid. If a gas were used, the system would then become pneumatic.

Why is a liquid used in a hydraulic system?

For all practical purposes, liquids are noncompres-sible. The compressibility of a fluid is its ability to occupy a smaller space, or volume, than it did before it was compressed. Because of this fact, the hydraulic system is almost instantaneous in action. The liquid in a tube acts like a solid bar of steel. If the liquid at one end of the tube is moved by means of a piston, a piston at the other end will immediately respond, as shown in figure 3. The motion of the second piston will be as smooth or as rough as was the motion of the first piston.

FORCE AND PRESSURE

In our study of hydraulics, we will have the occasion to use two highly important words very often—Force and Pressure. It is essential that the meanings of these two terms be thoroughly understood and well remembered, for they play vital roles in aircraft hydraulics.

FORCE is usually defined as a push or pull. The impor-

Figure 3.—Fluid action on a simple piston.

tant thing to remember is that force is independent of the area of surface to which it is applied. This means that a force of any intensity, however great or small, may be applied to an object of any size, however great or small.

For example, a 20-pound force may be exerted in pushing a thumbtack (a very small object) into a bulletin board, or in pulling a table (a very large object) across a room.

Pressure is the amount of force distributed over each unit of area. This tells us that when using pressure, we are considering the area to which the force is applied.

Pressure may be designated as so many pounds per square centimeter, per square foot, per square yard, and so on. The standard unit used in hydraulics to designate pressure is pounds per square inch, (p. s. i.).

It is important to bear in mind that, when referring to pressure, we deal with the amount of force acting upon one square inch of area.

Computing Force, Pressure, and Area

A simplified form of writing a long statement or sentence with the use of symbols is known as a formula. For example, rather than say: "Length times width equals area," we simply write:

L X W = A

When dealing with pressure, force, or area, we use a similar method of shorthand.

TO FIND FORCE:

It is known that force equals pressure times area.

Substituting F for force, P for pressure, and A for area, we have:

F = P X A

To find pressure:

It has been found that pressure equals force divided by area. For the sake of brevity and simplicity, let us condense this statement into symbols:

TO FIND AREA :

Since area equals force divided by pressure, we merely state:

$$A = \frac{F}{P}$$

Because these formulas are so often used, a trick device to aid in remembering them has been developed. In figure 4, any letter may be expressed as the product or quotient of the other two, depending upon their positions in the diagram.

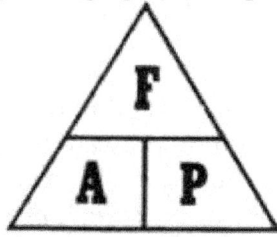

Figure 4.—Device for determining force, area, and pressure formulas.

PASCAL'S PRINCIPLE

In 1612, a Frenchman named Pascal discovered that the pressure exerted on any part of a confined liquid is transmitted to all parts of the liquid, regardless of the shape of the container. Now let us apply this knowledge as we might expect to find it used in hydraulics, and in so doing check our understanding of the difference between force and pressure as well as our understanding of Pascal's principle, or law.

In figure 5 is shown a cylinder filled with liquid upon which a piston is being forced with a 100-pound weight.

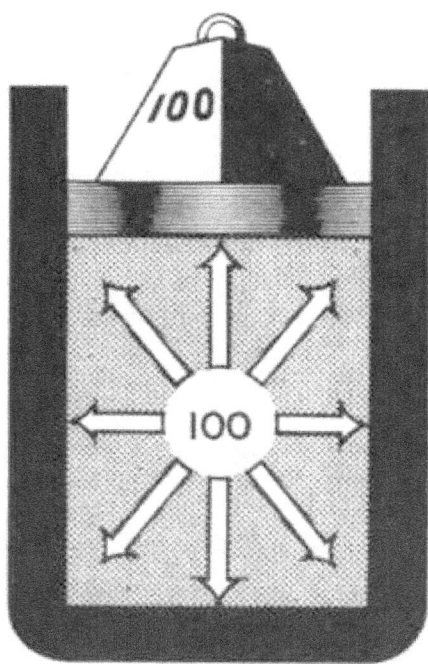

Figure 5.—Illustration of the equal distribution of force.

Assuming that we do not know the area of the piston, will it be possible to determine the pressure acting upon the liquid?

Yes, providing a pressure gage can be connected to the cylinder. Otherwise, such determination would not be possible. Since pressure takes into consideration definite areas, we could say only that a force of 100 pounds was acting upon the liquid.

However, if we knew that the area—not the diameter

—of the piston was 1 square inch, we could readily say that the pressure acting upon the liquid was 100 p. s. i. To get this answer, we used our formula for finding pressure:

~ 100 (force) , AA
P = 1 (area) = 100 P ' S ' 1 '

But suppose that the area of the piston is two square inches? What then would be the pressure acting on the liquid?

Using our formula, P = F/A, we write: P = 100/2, and get 50 p. s. i.

At first glance, it appears that the pressure should be 200 p. s. i., but it must be remembered that the entire surface of the piston is pressing down uniformly upon the liquid. Therefore, the 100 pounds of pressure is evenly divided over the surface of the liquid upon which it presses. Each of these two square inches will thus receive one-half of the weight or force, which will be 50 pounds. Since pressure means force exerted upon one square inch, the pressure is 50 p. s. i.

Let's add another and larger cylinder to the one considered in figure 6, and connect the two cylinders with a length of pipe. The condition pictured in figure 6 now exists.

In figure 6 is shown a 1 square-inch cylinder with a piston connected to a 3 square-inch cylinder and piston. If a force of 100 pounds is applied to a 1 square-inch piston, then we know—from our formula for pressure— that the pressure is 100 p. s. i. If we wish to determine the force acting upon the 3 square-inch piston, we use the formula for force, and find that the total force exerted upon the large area is 300 pounds.

The question now arises: If energy cannot be created or destroyed, how is it possible to exert 100 pounds of force and acquire 300 pounds of force?

It is true that energy cannot be destroyed or created, but it is possible to change energy. We can achieve an

Figure 6.—Force equals pressure til

increase in force if we are willing to sacrifice some other factor. In this instance, we sacrifice distance. Let us see what takes place.

Refer again to figure 6. If we were to push the small piston down one inch, we would displace one cubic inch of fluid. This fluid must go somewhere, and its only means of exit is through the connecting pipe into the large cylinder. This extra fluid is going to raise the large

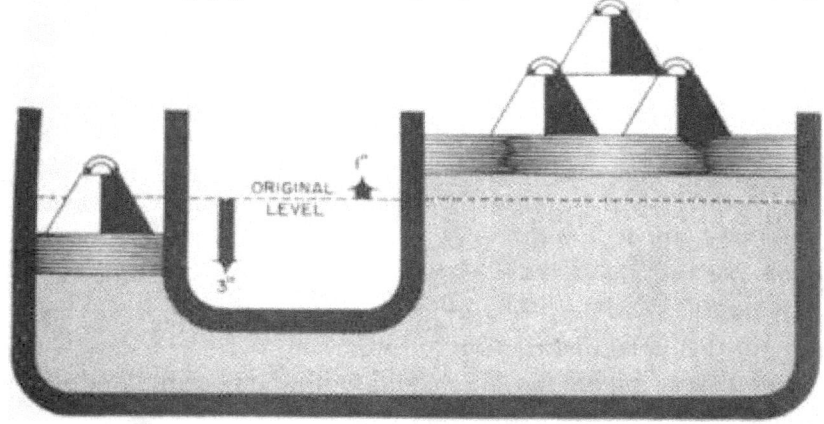

Figure 7.-Distance lost equals force gained.

piston, and our problem is to determine how high that piston will be raised.

The one cubic inch of fluid displaced by the small piston enters the large cylinder, and in that larger area it has three times its former space to occupy, as may be seen in figure 7.

If we were to raise the large piston 1 inch, 3 cubic inches of fluid would be required, but since the 1 cubic inch of fluid is distributed over the 3-inch area, each inch of that area will receive one-third cubic inch of fluid. Therefore, the large piston will rise only one-third inch.

APPLICATION OF BASIC PRINCIPLES TO AN AIRCRAFT HYDRAULIC SYSTEM

In an aircraft hydraulic system, the fluid confined in the tubing is used to transmit pressure from one location to another, and under static conditions, the pressure is the same at both ends of the tubing. This holds true regardless of the number of bends in the tubing or the length of the tubing.

When the tubing is connected to a cylinder whose area is greater than that of the tubing, and a piston is inserted in the cylinder, the force applied at the end of the tubing may be increased. Fluid forced into the tubing of the system at a certain rate moves the piston at a slower rate of speed but with much greater force. The motion of the piston and the force acting on it are transmitted by means of a piston rod to any mechanism that is to be operated on the airplane.

In figure 8 is illustrated a basic hydraulic system, showing two units cornected by a tube. The unit on the left (A) is the master cylinder, or hand pump, while that on the right, designated by (B), is the actuating cylinder. These units will be discussed in individual chapters later on in our study.

In the illustration in figure 8, there is a hydraulic fluid to the left of piston (C) in the hand pump, and below piston (D) in the actuating cylinder. The tube

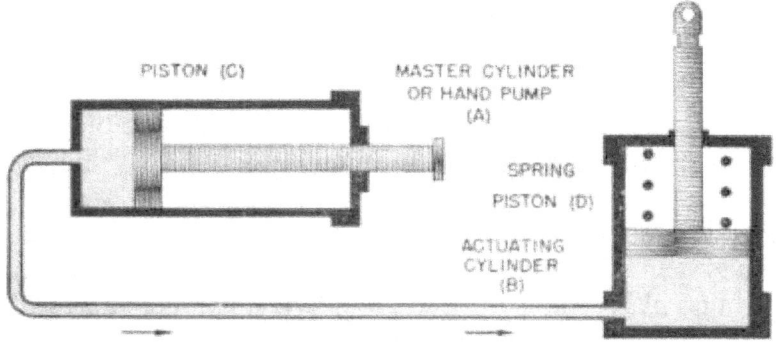

Figure 8.—Basic hydraulic system.

connecting the two units is also filled with the fluid.

When piston (C) in the hand pump moves to the left, it forces the fluid to flow into the lower part of the actuating cylinder (B). We learned that fluid is incompressible, and therefore piston (D) is forced upward, carrying the piston rod with it, and compressing the spring. When the pressure exerted by the hand pump is released, the compressed spring in the actuating cylinder returns piston (D) to its original position. Piston (C) also returns to its original position.

We called the unit at the left a master cylinder because this unit transforms force into pressure. We also may call it a hand pump, since a hand pump does the same thing.

In later chapters, we will discuss the necessary additions which must be added to the basic hydraulic system shown in figure 8, to cause it to function in an airplane. These conditions will include the reservoir, or source of hydraulic fluid, needed to replenish the fluid driven out of the master cylinder, or hand pump. Check valves will also be added to permit the fluid to flow in one direction but not in the other. Selector valves are other vital adjuncts. These are devices by means of which the flow of fluid may be directed in any of several directions. Power pumps and relief valves, pressure regu-

lators, accumulators, and many other devices will be fully discussed.

Having learned of the principles of hydraulics, let us now consider the primary element of the hydraulic system—the hydraulic fluid and the lines and tubing through which it flows to operate the many vital parts of the Navy's air fleets.

QUIZ

1. How does a liquid react when heated?

2. Why are liquids used in hydraulic systems?

3. That term which may be defined as push or pull and which is independent of the area of surface to which it is applied is known as what?

4. How does pressure differ from force?

5. State the formula for finding:

a. Area.

b. Pressure.

c. Force.

6. State the principle discovered by, Pascal.

7. Can energy be destroyed?

8. How much pressure is lost in the transmission of force in a hydraulic system from Master cylinder to actuating cylinder assuming a 60-inch length of tubing is used between the two cylinders?

CHAPTER 3

FLUIDS, SEALS, AND TUBING HYDRAULIC FLUID

Hydraulic fluid is the life-blood of the system through which it flows. It is the means of transmitting energy from the hydraulic heart—the pump—to the hydraulic muscles—the actuating cylinders.

The majority of naval aircraft use Air Force-Navy standard mineral-base fluid. However, some of the older transport and commercial type aircraft still use a vegetable-base fluid. It is most important that a clear distinction between the two types be established.

An extremely important point to remember is that the two oils are not interchangeable. Under no circumstance should a system be serviced with a type of fluid different from that specified for its use on its instruction plate. And since certain individual parts of a hydraulic system require vegetable-based fluids while others demand fluid

with mineral bases, the filling specifications of various units such as master cylinders, brake systems, shock struts, and shimmy dampers should be especially checked.

Comparisons Between Vegetable and Mineral Oils

Now let us consider the differences between the two types of hydraulic fluid. Table 1 furnishes comparisons between vegetable and mineral fluids, and should be carefully studied.

If the positive identity of the oil in a container or can is not established by the specification numbers listed in table 1 printed or stamped on the can, do not use that OIL in any hydraulic system.

The third comparison in table 1 indicates that

Table 1.— Hydraulic fluid chart.

different types of packings are used in hydraulic systems. Vegetable and mineral fluids must never be mixed because of the resultant damage to packings. ill case such error inadvertently occurs, the system must immediately be flushed, as described below, and all seals, packings, and hose lines replaced and the system refilled with new hydraulic fluid.

Cleanliness of the fluid is the factor that determines the efficient operation of the units in the system. Dust particles, metal chips, or other impurities in the fluid damage seals, and frequently cause hydraulic equipment to become inoperative. A great percent of troubles in the system are eliminated by keeping the fluid clean at all times.

Hydraulic fluid containers, disconnected lines, or the reservoir must not remain open any longer than necessary, so that dust and grit do not enter the oil. Fluid is

kept clean by filtering out foreign matter. It can then be reused. Filtering is accomplished by flowing the fluid through an external filter equipped with a micronic type filtering element of a finer grade than that used in the system.

In the case of obvious contamination of the fluid, flushing of the entire system is a necessary procedure.

SEALS AND PACKINGS

Seals and packings are as vital to hydraulic fluid as hydraulic fluid is essential to seals and packings for the maintenance of pressure in the hydraulic system. One is not complete without the other.

The Hydraulic Seal Chart (see table 2) describes several types of seals and packings, the purpose and installation of which will be discussed in detail later in the chapter.

Hydraulic Seals

Hydraulic seals, sometimes referred to as packings, are designed for a twofold purpose. Their function is to seal pressure in as well as to keep air out of hydraulic lines. Basically, all seal designs may be classified as one of four general types.

The chevron seal (V-rings) is manufactured in two forms—natural or synthetic rubber.

Seals of this type are used on some models of shock struts, and generally are V-shaped, always requiring reinforcement. The point of the V on this seal is supported by a female packing former, and the open end of the V has a male packing former as support. These formers are made of fiber, plastic, or metal.

The V-rings listed below are approved for use in hydraulic system units and shock struts using Specification No. AN-0-366 hydraulic fluid.

The V-rings made from Garlock Compound No. 7815 must be used only in hydraulic system units and struts

that specify use of AAF Specification No. 3586 (blue color) fluid. All procurements of packing rings made from Garlock Compound No. 7815 for use in Specification No. 3586 fluid will have a blue color band or blue dot on the heel of the packing ring.

When installing chevron-type seals, it is important to ascertain that the lips, or open end of the V, face toward the pressure or flow. Some formers used in shock struts are quite thick and fit into the cylinders with very little clearance. Care should be taken not to bind the formers inside the strut, as some difficulty—resulting in damage to the former or scratching of the finely-machined surface of the strut—may be experienced in removing them.

When installing chevron seals over threads, shimstock should be wrapped over the threads to protect the lips of the seals. Another point to bear in mind when installing a series of chevron seals is to properly "seat" the seals on the formers and upon each other. Never force seals into place, but make certain that they fit properly.

Cup seals are made both of natural and synthetic rubber, and are used on various units such as selector valves and brakes. Cup seals sometimes are used to seal against dirt and air as well as against fluid. An important point to remember is that cup seals are effective in one direction only. Where required, shimstock should be used to protect the lips of the seals during installation.

O-ring packings, the most commonly used type of seal, are used to prevent internal and external leakage in

stationary and moving parts. O-ring seals are doughnut-shaped and are used for effective sealing in both directions. Like cup seals, they are made of either natural or synthetic rubber, and are used in many units such as piston heads and selector valve sleeves.

The sealing characteristics of O-ring seals is their ability to spread against the groove in which they are installed when pressure is applied on one side or the other of the working part.

In all systems having pressures higher than 1,500 p. s. i., leather back-up rings are used to protect the

Table 4.— O-ring seals.

The above lists are subject to frequent change. Hydraulic packing and gaskets listed in Qualified Products List (QPL-5516-1 or any revision thereto) have been qualified under the requirements for the product as specified in the latest effective issue of the applicable specification (AN-P-79) and drawings (AN-902, AN-62-30, AN-62-25, AN-62-38, and AN-62-27) for use on Air Force and Navy aircraft.

"Military Qualified Products List of Products Qualified under Military Specification" can be obtained from the Commanding Officer, U. S. Naval Air Station, Johnsville, Pa.

PISTON "0"RING SEAL

X /

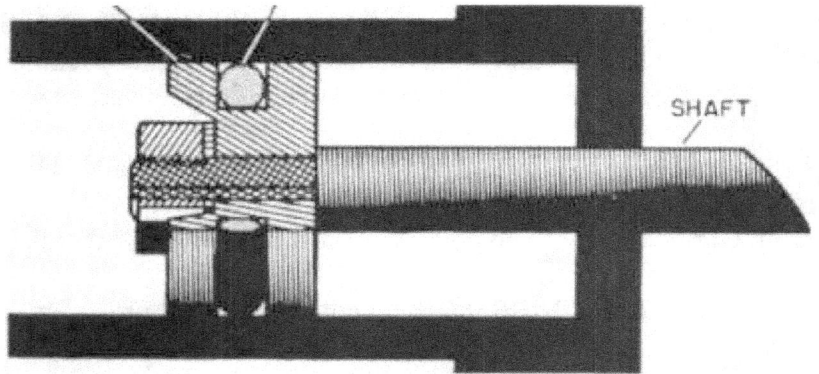

SHAFT

Figure 9.-—O-ring seals installed on a piston.

O-rings. For example, actuating cylinders in the model AM-1 airplane that are subjected to high or excessive pressure are equipped with leather back-up rings on one or both sides of the O-ring packings in the cylinders. In the wing folding actuating cylinder, where there is the possibility of excessive amount of cylinder barrel stretch, due to high pressure, leather back-up rings are used in the piston. The prime function of the leather back-up ring is to prevent extrusion of the doughnut seal under high or excessive pressure; thus the O-ring packing is protected against damage or wear.

The O-rings listed in table 4 are approved for use in hydraulic system units using Specification No. AN-0-366 (red color) hydraulic fluid.

In addition to using shimstock as a precaution while installing O-ring seals, it is best to fit this type of packing into place by "rolling" it over an object called a bullet. Bullets, or tapered plugs, may be devised to correspond with the diameter of each piston or unit where an O-ring is installed. Its tapered nose enables the seal to be rolled on and easily located in the groove of the piston. O-rings should always be sealed in hydraulic liquid before installation.

It is highly important that seals specified by the manufacturer of a unit be used to replace O-rings.

Metallic seals, more commonly known as crush washers, are designed for high-pressure use as air valve seats on accumulators and shock struts, and for seats on fittings which screw into units. This type of seal, used on non-moving parts, is generally made of soft aluminum.

The crush washer seal is installed flat. In fittings designed for this type of seal, the seat of the seal is grooved to provide additional sealing. These grooves must be perfectly clean before installation, and should be carefully checked for scores or scratches. Figure 10 shows the manner in which a crush washer is fitted into a unit.

Although square seals may be found in certain units, they are being replaced with O-rings. However, the same care must be used in identification and installation as is exercised with other types of seals.

Lapped Surfaces

Another method of sealing, which although not especially suited to replacement is worthy of mention here, is known as lapped surfaces. This is an operation requiring great care and judgment.

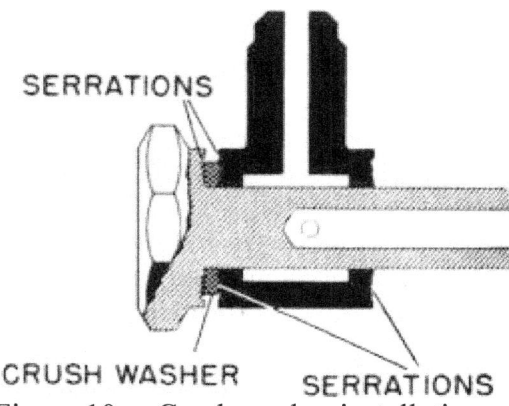

SERRATIONS

CRUSH WASHER   SERRATIONS

Figure 10.—Crush washer installation.

Lapped surfaces usually come in mated pairs which have been ground and polished so as to fit each other snugly. They should be carefully arranged upon disassembly of a unit to avoid mixing. Because of the skillful hand work involved in their manufacture, they are extremely costly and should be handled with extreme care.

Some lapped surfaces will be found in high-pressure hydraulic units having such tolerances as 0.0001-inch to 0.0003-inch, and should be handled at normal temperatures with clean, oily hands—the "oily" being clean hydraulic fluid.

Extreme care is absolutely necessary in unit disassembly and assembly. As units are disassembled, the parts should be submerged in AN-0-7 oil and left there until reassembled. Working with all units completely submerged in clean hydraulic fluid at normal temperatures will keep the units at proper sizes, thus facilitating assembly. They should never be forced, since galling or separating them may injure them beyond repair.

HYDRAULIC TUBING

In the human body, an intricate system of arteries and veins transports blood to and from the various organs. In the hydraulic system, tubing may be likened to the veins and arteries because of its function of carrying fluid from unit to unit. Fittings are used to connect tubing to hydraulic units and to join lengths of tubing together.

This discussion, therefore, will consider the two types of tubing used in hydraulic systems— rigid tubing and

FLEXIBLE HOSE.

Rigid Tubing

Rigid tubing for aircraft hydraulic systems is made principally of three materials.

Aluminum alloy tubing (52SO) is prescribed for systems with operating pressures up to 1,500 p. s. i.

Aluminum Alloy Tubing (61ST) is used for systems with operating pressures up to 3000 p. s. i. Stainless steel tubing (18-8) is employed in systems where lines are subjected to heat.

Rigid tubing sizes are the outside diameter (OD), and are designed in sixteenth-inches. Thus, number eight tubing would have an outside diameter of % B -inch (or %-inch) OD.

Flexible Hose

Flexible hose is often used to carry fluid to and from actuating cylinders which are hinged to the plane structure. It is used on the wing fold of carrier aircraft and wherever excessive vibration is encountered.

Flexible hose is constructed of seamless synthetic rubber inner-tubing and wire or cotton braids impregnated with synthetic rubber. Sizes of flexible hose are designated in sixteenth-inch inside diameter (ID), and are identical with the corresponding sizes of rigid tubing with which it can be used.

Low PRESSURE HOSE (Specification AN-H-29) is constructed of three plies of material. It consists of a thin-walled synthetic rubber tube, one braided fabric ply, and a thin rubber cover. This hose is identified by one broken yellow stripe running its length. This hose is used only with detachable end fittings in systems containing fuel, engine oil, hydraulic fluid, alcohol, water, and air, where the pressure does not exceed 300 p. s. i.

Medium pressure hose (Specification AN-H-24). This fire-resistant hose is constructed of thin synthetic rubber tube covered with one fabric braid, one wire braid, and an outer braided fabric cover. This hose is recommended for all flexible connections forward of the firewall. Identification is made by double yellow stripes.

High pressure hose (Specification AN-H-28). This is a flexible hose recommended for 3,000 p. s. i. and is available only in assemblies. The composition consists of a synthetic rubber tube with two wire braids covered

with fabric braid on a thin rubber cover, with attached end fittings. This hose is marked with a single broken yellow stripe.

HYDRAULIC LINE COLOR CODE

Hydraulic lines are plainly marked with color bands spaced to facilitate rapid identification. The color code alternates in one-half inch blue and yellow bands. One blue-yellow-blue band should be conspicuous on each section of tubing 24 inches or less in length, if both ends are in the same compartment.

On longer sections of tubing, the color bands should be located at each end. Where one length of tubing passes through more than one compartment, or bulkhead, additional bands should be used so that at least one band is visible in each compartment, one on either side of a bulkhead.

Emergency air pressure lines are coded with light green and yellow.

FLARING TUBING ENDS

Tube flaring is an art requiring a great deal of precision. Many system failures can be traced directly to poor flares. Whenever power flaring machines are available, the tubing ends should be squared, and flared on the machine. When power flaring machines are not available, the following steps should be observed in flaring tube ends.

After cutting the rough sections of tubing to estimated lengths, bend the tubing to proper shape, and cut to correct size.

To insure a good flared joint, the tube must be cut off square. This can be accomplished by the use of the tube cutter (see figure 11).

In operating the tube cutter, do not feed the wheel too rapidly. Moderate or light tension on the ball will maintain even tension on the cutter wheel and prevent deformation or excessive burr on soft tubes. If a tube

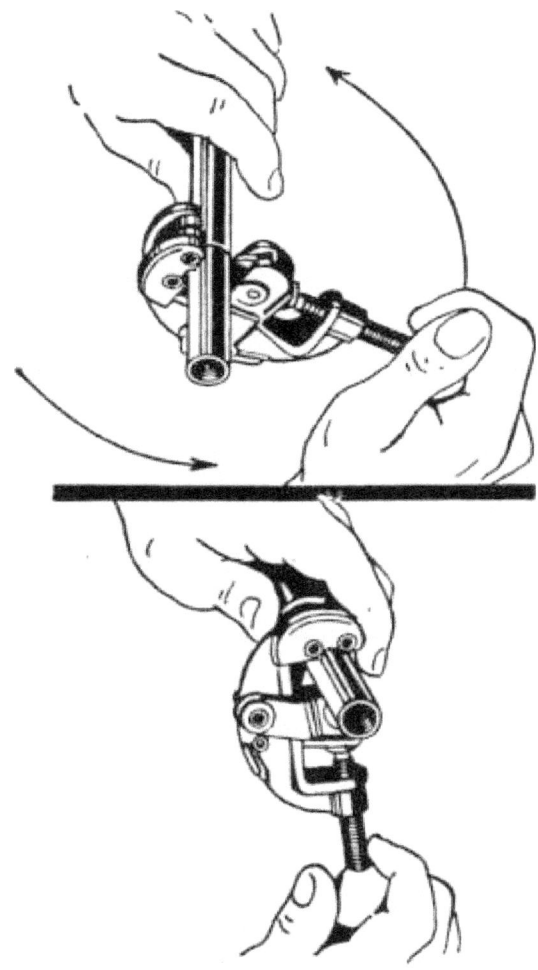

Figure 11.-Tube cutter.

cutter is not available, square off the ends with a fine-cut mill file. Remove all filings, chips, burrs, and grit from the inside of the tube to avoid pock marks or scratches on the inside surface of the flare. Use of compressed air, if it is available, blown through the tube provides the best method of cleaning copper or aluminum tubes. This done, insert nuts and sleeves on the tubing

and flare the tube ends with a flaring tool such as the Parker type or its equivalent.

The completed flare should retain at least 82 percent of its original wall thickness. Care should therefore be taken not to thin out the flare too much.

The outside edge of the flare must be in line with the outer edge of the sleeve in order to guard against leakage and to insure correct installation of the fittings.

The flare grip-die, designed to grip without damaging the tubing, consists of two steel blocks placed side by side and held in alignment by three steel pilot pins pressed into one block and extending into corresponding holes in the other block. A number of countersunk holes,

Figure 12.—Parker-type flaring tool.

33

varying in size to conform with tube diameters, are drilled along the grip-die section of the tool. The flaring tool consists of a cylindrical bar tapered at one end to correspond with the angles of the countersunk holes in the die.

To make a flared tube-end, two nuts and two sleeves should first be slipped over the tube ends, after which the tube is inserted into the grip-die section through the space provided when the two sections of the tool are parted. The tube end should be approximately %4-inch above the surface of the grip-die. Secure the assembly, with the tapered end of the flaring tool in the tube end, and tap lightly with a hammer—while rotating the tapered end—until the walls of the tube take the shape of the countersunk hole in the die.

In an emergency, a satisfactory flare may be made in the manner outlined below.

Drill a hole in a wooden block slightly smaller in diameter than the sleeve or line. Split the block lengthwise through the drilled hole, then clamp the block over the line with a vise or C-clamp. Using the sleeve as a form,

Figure 13.—Flared grip-die.

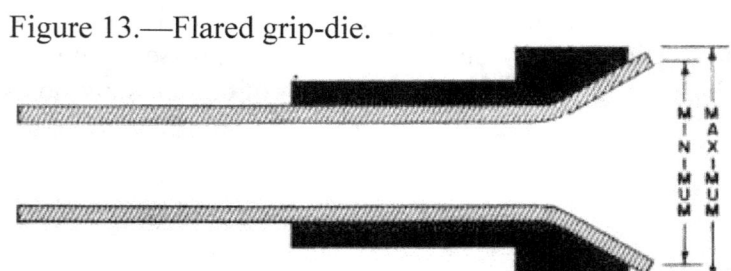

Figure 14.—Proper dimension of a flare.

flare the tube to the same angle as the sleeve. A blunt instrument, such as a punch or a bolt filed down to the correct flare angle, may be used. Square off the ends, remove all burrs, and polish the rough edges.

The flare should not extend more than ^«-inch beyond the tip of the sleeve. Figure 14 illustrates the proper dimension of a flare.

The interior of all tubing should be thoroughly blown out, flushed, or cleaned otherwise just before installation in the airplane.

All flares should be carefully examined for minor cracks before installation of the line.

TUBE BENDING

The objective in tube bending is to obtain a smooth bend without flattening the tube. Tubing should always be bent with the aid of mechanical tube-benders. These instruments are designed in specific sizes and bend radii for each size of tubing. The correct bend radius is three times the outside diameter of the tubing. If the tube bending must be done by hand, the minimum bend radius should be six times the outside diameter.

Tubing must be bent to the exact angle. A tube which springs out of position when disconnected from a fitting has not been properly bent and is under mechanical stress.

A bend radius must not begin within a length equal to one-half of the diameter of the tube from the end of the

sleeve. A bend commencing within that area will bind the sleeve against the flare and cut the flare on one side when tightened.

INSTALLATION OF LINES

In replacing hydraulic lines, always make certain that the replacement tubing is of the same type of metal and size, and that it has been formed to the same shape as the original line being replaced.

The following six important rules should be followed when installing hydraulic lines:

1. Check all flares for out-of-round and cracks.

2. Blow out all lines with clean, dry air.

3. Lubricate all male threads of fittings sparingly. For this purpose, use hydraulic fluid of the type employed in the particular airplane system.

4. Never scrape a flare over a mating fitting when aligning the line.

5. Make certain that the tube flare meets the fitting squarely and fully before starting the nut. Never draw the flare to the fitting with the nut, as this may cut the flare.

6. Never tighten the nut when the line is under tension.

HANDLING TUBING

Tubing requires careful storage and handling to prevent nicking, scratching, and denting. Stack tubing only where it will not be exposed to falling objects, and avoid handling mixed shapes of tubing. Such practice tends to scar or bend the tubing.

Do not hang lamp cords or other objects from hydraulic lines, and never climb upon or pull against tubing.

Any dent affecting less than 20 percent of the tube diameter is not objectionable unless such damage occurs on the heel of a sharp bend radius. If a nick or scratch in aluminum alloy tubing penetrates no deeper than 10 percent of the wall thickness, the tubing may be reworked

by burnishing. Use a fine-grade emery cloth and oil for this purpose, then finish and polish with crocus cloth and oil.

FLEXIBLE HOSE REPLACEMENT

When replacing flexible hose, you may re-use the end fittings from the section removed if they are not damaged. New hose should be of the same size, type, and length as the section being replaced. The following procedure should be followed in replacing flexible hose.

Wet the blade of a sharp knife in oil or water and cut the hose squarely. Clean the inner surface of the hose, and place the socket in the assembly-block and tighten the block in a vise. Screw the hose counterclockwise (left-hand threads) into the socket until the hose rests on the shoulder of the socket. Then back off one-fourth turn.

Apply hydraulic fluid to the inside of the hose, then insert a mandrel in the hose and apply a flaring motion. Work the mandrel in and out until no binding is evident. Install a male-type nipple in the mandrel.

Figure 15.—Mandrel used for flaring tubing.

Dip the nipple and mandrel in fluid, then insert both in the socket and apply pressure upon the knob of the tool with the palm of your hand. Screw the nipple into the socket by using a wrench on the hex of the nipple.

When installing flexible hose, be certain that the hose is not excessively twisted or bent. Such treatment will considerably reduce its life. The amount of twist can be determined from the white line running along the length of the hose. This line should not spiral around the hose.

Flexible hose should be protected from chafing by a light wrapping of tape, but only where necessary. Where
a flexible hose has a bulkhead-end fitting, a washer should be installed only beneath the lock-nut.

The minimum bend radius for flexible hose is nine times the outside diameter. By "bend radius" is meant the radius of any curve to a part which is bent in forming. Sharper bends will reduce the bursting pressure of the hose to 80 percent of its rated value. Hose should be installed so that it will be subjected to a minimum of flexing during operation.

A hose must never be stretched tightly between two fittings. About 5 percent to 8 percent of its total length must be allowed as slack to provide freedom of operation under pressure.

And finally, don't forget to inspect hose regularly for cuts and abrasions.

FLEXIBLE HOSE ASSEMBLY

Select the hose, fittings, and assembly tools as shown in figure 16. The tools required are wrenches of correct size, vise, assembly block, and an oil can.

After obtaining the hose, to get the required length, subtract dimension "P" from the length of the assembly, as shown in table 5. Cut the hose to the proper length, then using a vise and assembly block, prepare for assembly.

Table 5.— Length of assembly
ASSEMBLY BLOCK

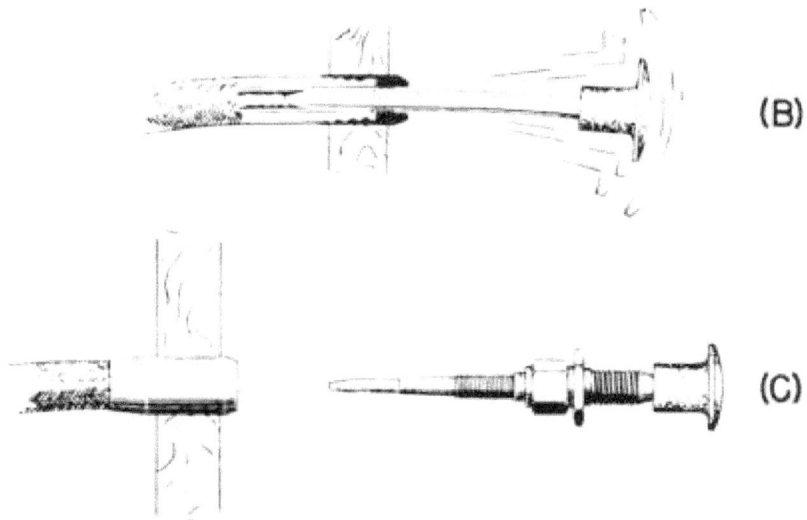

Figure 16.—Flexible hose equipment and assembly.

After screwing the hose into the socket (socket has left-hand threads) back off a quarter turn to prevent binding during assembly. Oil the assembly tool with hydraulic fluid, and insert into the hose. Use a flaring

motion, working the tool in and out to enlarge the hole in the hose, as shown in (B) of figure 16. Install the nipple on the assembly tool and tighten with a wrench, as may be seen in (C).

Apply oil on the threaded end of the nipple and to the end of the hose and socket assembly, then screw the nipple into the socket. Screw the swivel-type nipples in until there is a clearance of % 4 -inch to y.^-inch between back of nut and socket, as shown in (E). Screw all male nipples up snug against the face of the socket.

FITTINGS

There are two types of AN standard tube connectors, two-piece and one-piece. The TWO-PIECE CONNECTOR, consisting of a sleeve and nut, is the one most commonly used.

The sleeve of the two-piece connector fits directly over the tubing, and one end is countersunk at the same angle as the tubing-flare. The nut fits over the sleeve and, when tightened, draws the sleeve and tubing-flare tightly against the male fitting to form a seal. This occurs because the male fitting has a beveled surface with the same angle as the inside of the flare.

The twofold purpose of the sleeve is to support the tubing so that vibration does not concentrate at the edge of the flare, and to distribute the shearing action over a wider area for added strength.

The one-piece tube connector has no sleeve. The inside of the nut fits against the outside of the tubing-flare and clamps it against the male fitting.

Practically all of the fittings used in current aircraft are these two AN standard types. They are designed to mate with flared lines. They are made of aluminum alloy and also of steel. Aluminum AN fittings are blue in color; steel AN fittings are black.

When connecting fittings, always thread them together with your fingers. If the thread binds, examine the fit-

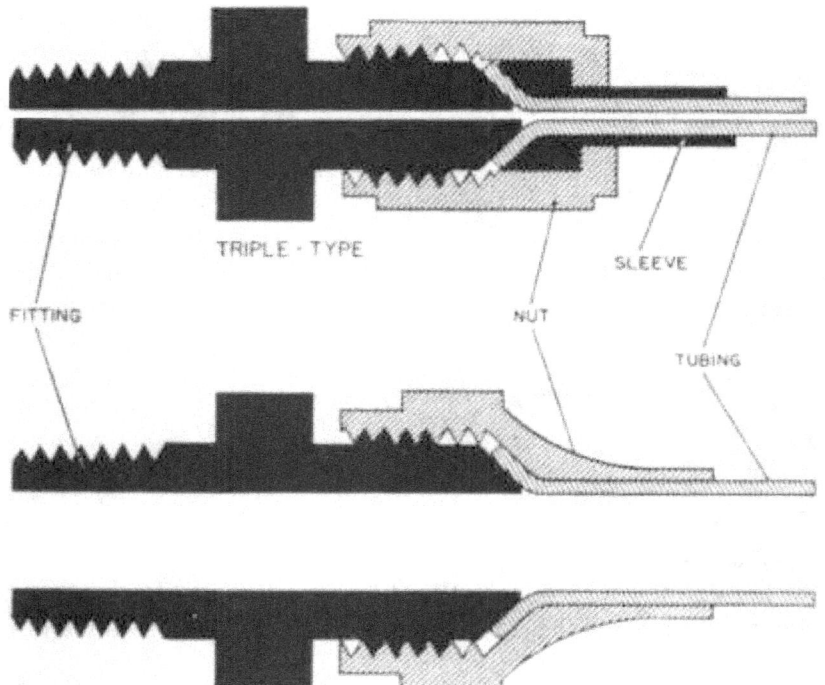

STANDARD

Figure 17.-Triple-type and standard tube connector*.

tings to make certain that the parts are clean and contain no imperfect threads. Also make certain that mating parts are properly aligned to avoid cross threading. Never force a binding thread. Additional turning will strip the threads.

Quick-Disconnect Fittings

When servicing hydraulic systems, it is often necessary to disconnect lines whose opening would permit large quantities of fluid to drain from the system. By using quick-disconnect fittings, it is possible to discon-

nect lines without this loss of fluid or entrance of air. Quick-disconnect couplings are installed at the firewall and at other locations where frequent disconnections are made.

The quick-disconnect coupling is made in two sections and joined by a union nut. One section consists of a body with a flange into which an end-adapter is screwed. Within this section is a poppet valve held on its seat bv a spring. The other half of a quick-disconnect fitting consists of a body in which a tubular valve is fixed and within which a sleeve slides on the outside diameter of the tubular valve. The seat of the sleeve is held against the head of the tubular valve by a spring. An O-ring packing provides a seal between the sleeve and the body.

When connecting the fittings, the seating portion of the male body makes contact with the sliding sleeve of the female body. At the same time, the head of the male body makes contact with the face of the poppet valve. As the union nut is tightened, the tubular valve pushes the poppet valve off its seat, and the male sleeve or flange pushes the female valve from its seat. Now the coupling is assembled, and permits free passage of fluid.

Coupling or uncoupling of quick-disconnecting fittings is to be done by hand only. Couplings are not designed to withstand forces resulting from coupling or uncoupling by wrenches.

K. -
QUIZ

1. How can you ascertain whether to use a vegetable-base or mineral-base hydraulic fluid in a particular airplane?

2. Do all parts of a hydraulic system use the same type of hydraulic fluid?

3. What difference might you observe in the color of mineral-base and vegetable-base hydraulic fluid?

4. What is the identifying odor of vegetable-base hydraulic fluid?

5. How must you identify hydraulic fluid before using it in any hydraulic system?

6. Why must vegetable and mineral fluids never be mixed in a hydraulic system?

7. In which direction should the open end of the V face when chevron-type seals are properly installed?

8. Of what material are crush washers made?

9. What two types of hydraulic tubing are most commonly used?

10. How may hydraulic lines be most rapidly identified?

11. What percent of its original wall thickness should a complete tubing flare retain?

12. What is the correct bend radius for a length of tubing when bending is to be done with the aid of a mechanical tube bender?

13. What are the primary details to be observed as a primary step in replacing hydraulic lines?

14. Why must you never draw the flare in a hydraulic line to the fitting with the nut?

15. What is the minimum bend radius for flexible hose?

16. What percent of a length of flexible hose must be allowed as slack to provide freedom of operation under pressure?

17. What are the two types of AN standard tube connectors?

18. What is the advantage of using quick-disconnect fittings in a hydraulic system?

19. Where in the hydraulic system on an airplane would you expect to find quick-disconnect couplings installed?

20. Should coupling and uncoupling of quick-disconnect fittings be done by hand or with a special wrench?

CHAPTER 4

RESERVOIRS AND FILTERS DESCRIPTION

In the hydraulic system, the reservoir is the tank in which fluid is stored. This unit functions both as the starting point and place of return for fluid circulating through the lines.

The primary purpose of a reservoir is to supply the operating needs of the hydraulic system and to replenish fluid lost through leakage. It also furnishes a place for the escape of air from the fluid; acts as a container for excess fluid forced out of the system by thermal expansion ; aids in cooling the fluid, and removes foreign matter with its filters and pumps.

Hydraulic reservoirs are usually located at the highest point in the system so that the force of gravity will keep the pumps primed.

Reservoirs are designed and constructed in varying shapes and sizes, but generally they are welded cylindrical tanks formed of welded aluminum alloy sheet.

Welded magnesium also has been successfully used. Most reservoirs are mounted on padded supports and held in place by felt-lined straps. The unit is usually mounted at the top of the forward firewall, or in one of the engine nacelles.

There are at least two ports in the bottom of the reservoir. The larger of these is attached to the engine-pump supply port. The other port is joined to the hand-pump suction line.

Projecting into the reservoir is a standpipe which is attached to the engine pump port at the bottom of the tank. The standpipe serves two purposes—it maintains a reserve supply of fluid for emergency hand-pump action, and it prevents foreign matter from reaching the engine pump by permitting such particles to settle to the bottom of the tank.

The fluid supply of the hand pump port comes directly from the bottom of the reservoir where it is filtered before entering the hand pump port. A drain line and shut-off valve is sometimes taken from the hand-pump supply line.

In some types of reservoirs, the main RETURN LINE from the system enters on the side and below the fluid level. This prevents "boiling" of the fluid and permits air bubbles to escape more readily. In other types of reservoirs, the return line enters at the top and the fluid reaches the main port of the reservoir through filters. We will discuss filters in detail later in this chapter.

Baffles and fins, installed within the reservoir, prevent excessive vortexing and surging of fluid, thus eliminating air from fluid entering the pump supply lines. By "vortexing" is meant a "Whirling" of the fluid within the reservoir tank.

Figure 18 illustrates a typical reservoir, portraying in detail the parts described above.

The filler opening usually is near the top of the reservoir, but in most cases it is installed sufficiently low on the reservoir tank to prevent over-filling. A finger

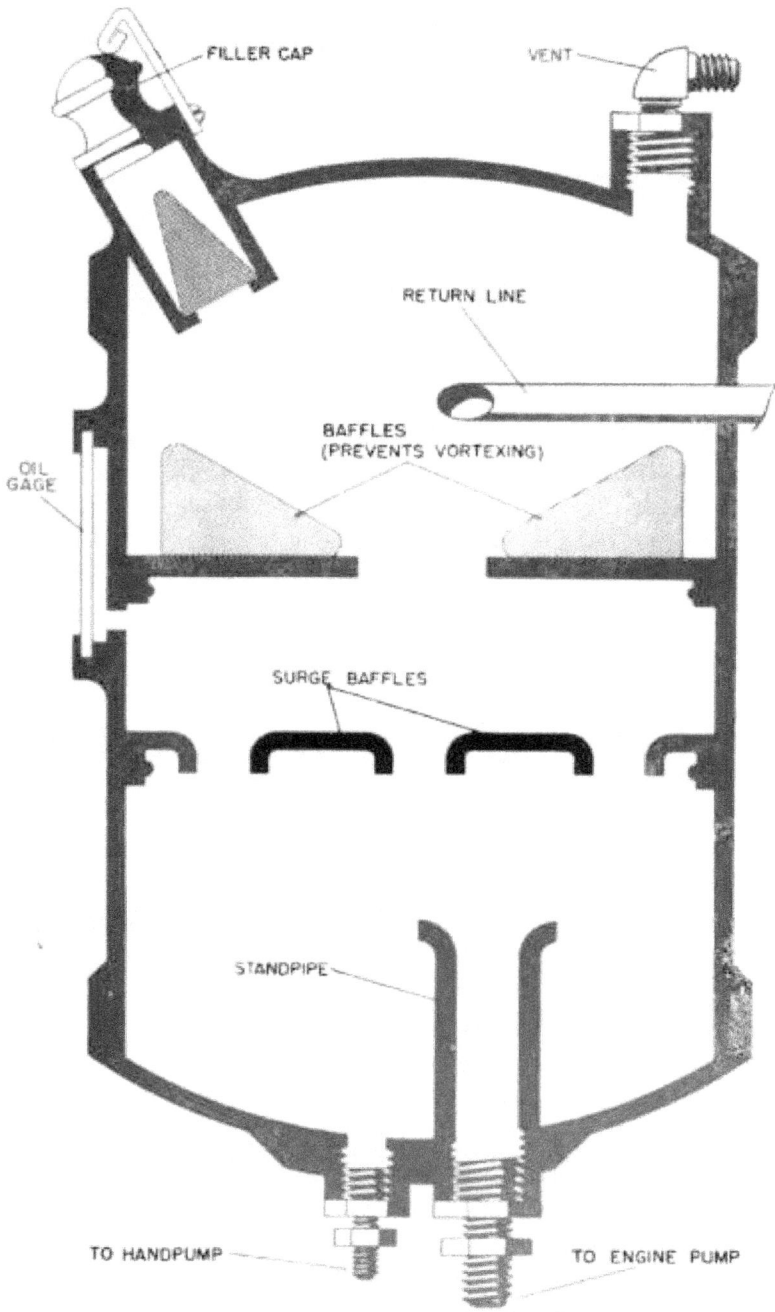

FILLER CAP

VENT

RETURN LINE

BAFFLES
(PREVENTS VORTEXING)

OIL
GAGE

SURGE BAFFLES

STANDPIPE

TO HANDPUMP

TO ENGINE PUMP

Figure 18.—Schematic diagram of a typical reservoir.

strainer is included in the filler opening to strain all fluid added to the system.

A VENT LINE, installed in the top of the reservoir, runs overboard at a point below the tank. This permits the reservoir to "breathe" as the fluid level varies. That is, when the fluid level rises, the air is forced out, and when the fluid level falls, air at atmospheric pressure occupies the space on top of the fluid level. Also, the vent makes it possible for any air that has entered the hydraulic system to find a means of escape. A check valve is placed in the vent line to prevent the loss of liquid while the plane is flying upside down.

A sight-gage to provide visual checking of fluid levels, or a dip-stick for measuring fluid levels, is installed on all reservoirs.

The capacity of a reservoir is determined for each airplane when the system is designed, and no general

RESERVOIR CAPACITY
TO RESERVOIR
FROM ATMOSPHERE

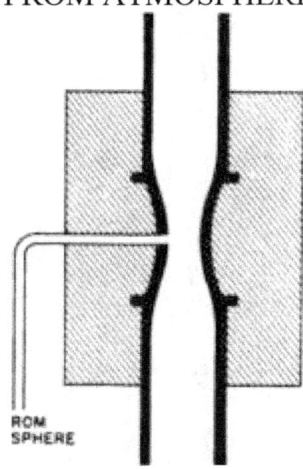

FROM EITHER PRESSURE MANIFOLD OR RETURN MANIFOLD

Figure 19.—Venturi fitting.

information can be given that will cover all cases. It is generally accepted, however, that the capacity should be large enough to provide for changes in volume caused by temperature changes and changes in the internal volume of all operating mechanisms.

The total amount of fluid required is provided at a point at least two inches higher than the top of the reservoir outlet that leads to the power-driven pump. Enough fluid must be provided below this outlet, but above the reservoir outlet which leads to the hand pump, to permit the operation of all mechanisms required to make a safe landing. This assures that no fluid is being returned to the reservoir during such operation.

Hydraulic systems in modern aircraft have total fluid capacities up to 18 gallons, the capacity of each system depending upon the number and size of its units.

PRESSURIZED RESERVOIRS

A reservoir may be pressurized by introducing air into the top of the reservoir. The purpose of pressurization is to insure a positive flow of fluid to the engine-driven pump during high-altitude flights where the atmospheric pressure is low. The decrease in pressure at high altitudes results in a lowered fluid supply to the engine-driven pump. This may starve the pump and cause its failure, because the pump depends upon the fluid passing through it for its own lubrication. Starvation of the pump will also cause the formation of injurious high-pressure peaks and pulsation in the system.

Figure 19 shows a diagrammatic drawing of a venturi PITTING installed in either a branch of the hydraulic-pressure manifold or the return line for the purpose of injecting air into a pressurized reservoir. Fluid under pressure is forced through the venturi tube. As the fluid passes the throat of the venturi, the pressure is lowered, thus causing the air to be sucked into the stream of hydraulic fluid. The mixture of air and fluid is then directed to the reservoir.

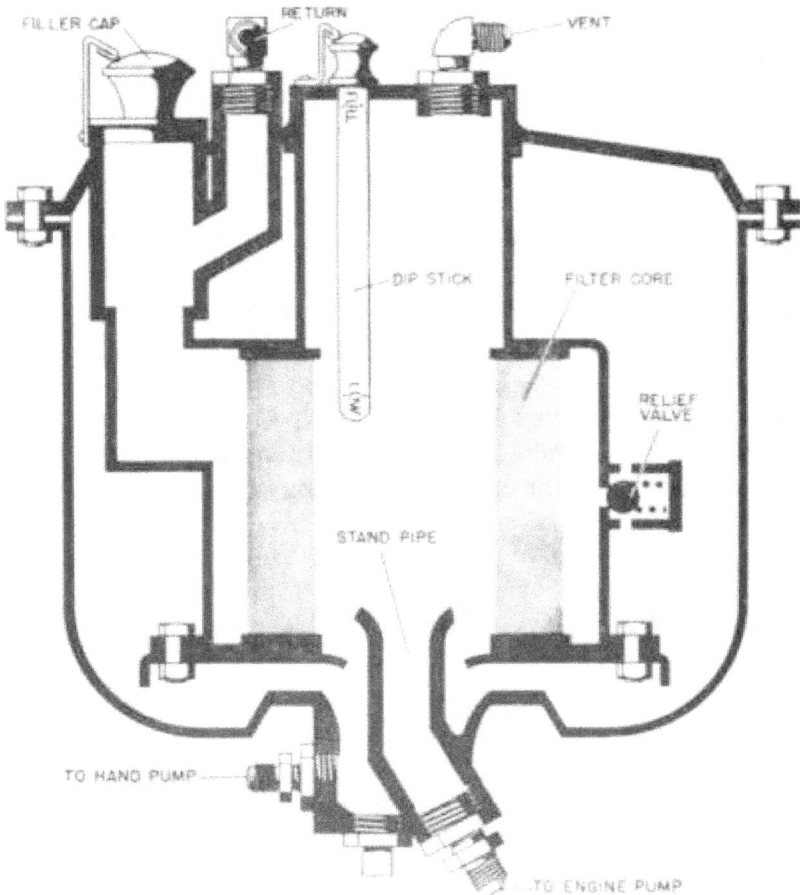

Figure 20.--Schematic diagram of a pretsurized reservoir.

There are two types of pressurized reservoirs—the closed type and the open type. The closed-type pressurized reservoir really contains two chambers, separated by a diaphragm or piston. The fluid is in one and air under pressure in the other. When fluid is needed in the lines, the compressed air forces it out of the reservoir. This type of reservoir is used in gun turrets and other small hydraulic systems.

The open-type pressurized reservoir has a relief

valve in the vent line set to relieve at a pressure a few pounds above atmospheric pressure. On some models, a small pump is used to force air into the reservoir at a pressure slightly below the relief-valve setting. This insures constant inlet pressure at the hydraulic pump, regardless of altitude. Figure 20 illustrates the pressurized reservoir.

FILTERS

A filter is a screening, or straining, device used to clean the hydraulic fluid, thus preventing foreign particles and contaminating substances from remaining in the system.

The hydraulic fluid holds in suspension tiny particles of metal that are deposited during the normal wear of the hydraulic units. These minute particles of metal may injure the units and parts through which they pass if they are not removed by a filter.

Filters may be located within the reservoir, in the pressure line, in the return line, or in any other location where they are needed to safeguard the hydraulic system against impurities in the fluid.

To perform efficiently the exacting duties of hydraulic maintenance, it is necessary that you understand the location, construction, and operation of filters used in the system.

Main system filters may be located in the pressure line, return line, or in the reservoir.

Classification

Filters may be grouped in two classes—those employing the principle of edge filtration, and filters which employ the principle of screening. The Cuno-type filter belongs in the first class, and the micronic-type filter belongs in the second class.

Of these, the micronic type has been found to be the most effective, and has been adopted as standard for

naval aircraft. Therefore, we will consider only the mi-cronic-type filter in our discussion.

Micronic-Type Filter

There are many variations of the micronic or absorption-type filters. In general, however, the construction, design of parts, and arrangement of this type of filter is determined by the amount of fluid in gallons-per-minute to pass through it for filtering. The filter is the replace-

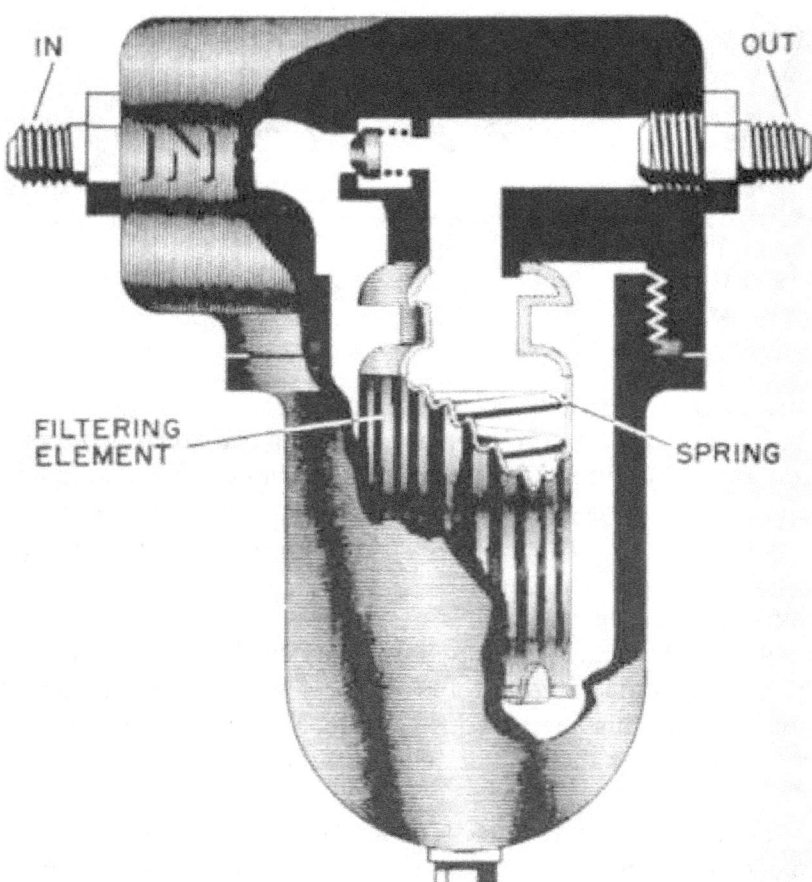

Figure 21 .-Micronic er ab«orption-typ« fllt.r.

ment element, consisting of a head assembly, filter element, and case.

The micronic-type filter was especially designed for hydraulic systems that require exceedingly close tolerances between working parts. The filtering element, or cartridge, is made of specially treated cellulose formed in convolutions (wrinkles) and designed so that solids greater than 10 microns (0.000394-inch) in size are removed.

Various models of this type of filter are made for conforming to the capacities of the hydraulic systems in which they are installed. In addition, they are made in types that are designed according to the place where they are to be installed. Thus there is one for the reservoir, another for the pressure and return lines, and still another type for the reservoir vent line. When a filter is placed in the reservoir vent line, its purpose is to filter dust and other contaminating substances from the air.

Figure 21 is an illustration of a micronic filter. This unit consists of a head containing the in-port and out-port, the case, and the filtering element.

A bypass check valve (not shown in the drawing) is located in the head. These parts are assembled into, one unit by inserting the filtering element in the case and screwing the case into the head. The seal prevents any leakage of fluid between the case and the head. The check valve is set to open and bypass the fluid directly from the in-port to the out-port if the filter becomes clogged.

Fluid enters through the in-port and must flow through the filtering element before it can discharge through the out-port.

At regular intervals, the case is unscrewed from the head and the filtering element is removed and discarded, and a new element inserted. Cleaning of clogged micronic-type elements is impractical, is apt to result in puncturing of the oil-soaked paper, and should be attempted only as a last resort. When replacement is not possible, plug or mask the outlet bushing in the removed

element to prevent entrance of dirt into the interior of the element. Clean the exterior of the element with proper solvent (see table 1).

Micronic filtering elements are used as reservoir filters in several late model aircraft. With this arrangement, no other main system filter is required.

Line Filters

Many types of line filters are used in various unit systems of the main hydraulic system. Line filters are smaller than main system filters, and generally are constructed with micronic-type filter elements. In some cases, line filters are used in conjunction with restrictors to prevent the orifice from becoming clogged.

Remember, the filtering action of all filters is from outside TO inside. This means that fluid to be filtered passes through all elements from outside through to the inside, leaving behind all foreign particles to large to pass through.

QUIZ

1. What is the primary purpose of the reservoir in a hydraulic system?

2. Where in the hydraulic system is the reservoir usually located?

3. What purposes does the standpipe serve?

4. What is the purpose of baffles and fins which are installed within the reservoir?

5. During high-altitude flights where atmospheric pressure is low, what device is used to assure a positive flow of hydraulic fluid to the engine-driven pump?

6. Which of the two types of pressurized reservoirs has a relief valve in the vent line set to relieve pressure at a few pounds above atmospheric pressure?

7. What principle of filtration is employed by the micronic-type filter?

8- Of what material is the filter cartridge of the micronic-type filter made?

9. What is the purpose of a filter placed in the reservoir vent line?

10. In what direction is all filtering action in hydraulic systems accomplished?

CHAPTER 5
HYDRAULIC PUMPS DEVELOPMENT

Pumps are the primary source of energy in aircraft hydraulic systems. They are the units that normally deliver hydraulic fluid under pressure for the actuating units. They convert the pumping force, or mechanical energy, into fluid pressure.

A pump, whether it be manual or engine-driven, serves the same purpose. A pump creates flow.

Let us trace the development of pumps from the hand pump to the intricate variable displacement unit. The latter, a piston-type mechanism, will be discussed in detail later in our study.

Early hand pumps worked exactly like the water pumps on the farm. In windmill water pumps, as the piston

moves downward a check valve opens to pick up the load, then closes to carry the load upward. The windmill device, known as a single-action pump, is quite satisfactory for supplying water for farm needs, but it proved wholly inadequate for aircraft use. It was too slow, and its flow was pulsating and intermittent.

A double-action pump that would create a flow with each piston stroke was developed for testing and emergency operation of aircraft units. In figure 22, a schematic drawing illustrates the fundamental difference between double-action and single-action pumps.

DOUBLE-ACTION PUMP OPERATION

In the drawing of the double-action pump, notice the CHECK valves at the intake port and in the piston.

When the piston of this pump is traveling outward, the check valve at the intake port opens to permit entrance of fluid into the cylinder. (Observe the amount of

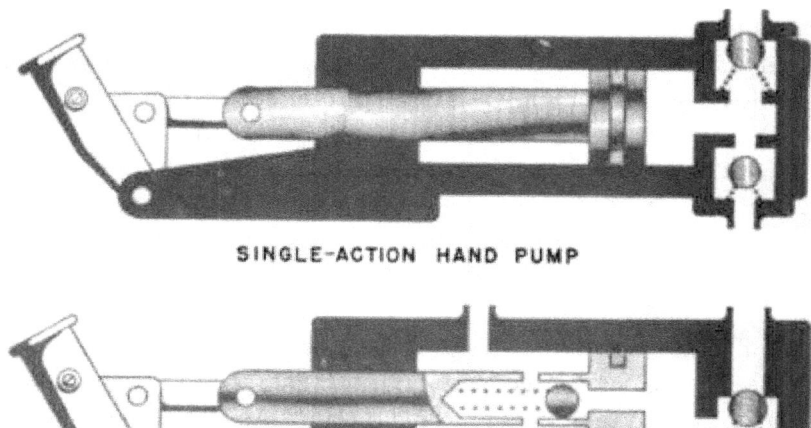

SINGLE-ACTION HAND PUMP

DOUBLE-ACTION HAND PUMP

Figure 22.—Schematic drawing showing the fundamental differences between single-action and double-action pumps.

CO

space provided for fluid when the piston is fully extended). When the piston travels upward, its check valve opens simultaneously with the closing of the check valve at the intake port. This action allows fluid from the large chamber to flow into the smaller compartment on the shaft side of the piston.

Now consider the difference in volume caused by the shaft as compared to that on the blank side of the piston. Inward travel of the piston displaces extra fluid, forcing it into the system. Therefore, fluid flows from the outlet port of the hand pump on both piston strokes. Notice also that the pressure, or outlet, port of the double-action pump is always next to the handle.

Hand pumps should be checked daily. Prolonged and effortless operation indicates internal leakage; spongy operation indicates the presence of air in the pump or system. Hand pumps are removed for overhaul at each engine change.

POWER PUMPS

Power pumps in general use today in aircraft hydraulic systems may be divided into two principal groups—constant displacement (constant delivery), and variable displacement (variable delivery).

The constant-displacement type of power pump has a fluid output that is constant for any rotational speed of the pump. For example, a pump may be designed to deliver three gallons per minute at 2,800 revolutions per minute (r. p. m.), regardless of the pressure demands made upon it, so long as the pressure demands do not exceed a designed maximum value. The F9F Panther plane uses Vickers 7-cylinder axial-type pumps in its hydraulic system. These pumps deliver nonpulsating flows with constant displacement per revolution. Maximum recommended pressure for continuous duty is 1,500 p. s. i. with a maximum recommended speed of 3,750 r. p. m.

The variable-displacement type of power pump has a fluid output that is varied to meet the demands of the

system. For example, a variable-displacement pump may be designed to maintain the pressure at 1,200 p. s. i. by-varying its fluid output from zero to seven gallons per minute.

GEAR PUMPS

Many types of power pumps (engine-driven pumps) have been designed to supply the high-flow and high-pressure requirements of various hydraulic systems. Our first consideration of such mechanisms will be the gear-type pump.

Basically, this type hydraulic pump consists of two counter-rotating meshed gears revolving in an aluminum-alloy housing which has a close-fitting cover. Figure 23 demonstrates the functioning of this unit.

In the schematic illustration, you will see that one gear, called the drive gear, is mounted on a drive shaft. The other, termed the idle gear, rotates on a plain shaft (a shaft without splines.)

The inport is connected to the reservoir, and the out-port is connected to the pressure line. The driving gear is attached to a drive shaft that extends outward from the housing, with seals to prevent leakage around the drive shaft.

When the driving gear turns in a counter-clockwise direction, it turns the driven gear in a clockwise direction. As the teeth pass the edge of the inport, oil is trapped between the teeth and the housing, and is then carried around the housing to the outport. As the teeth of the driving gear mesh with the teeth of the driven gear, the fluid between the teeth is displaced and is forced out of the outport and into the pressure line.

We have seen that the mating action of gears forces fluid into the system, and also that the gears are rotated by power supplied by the drive shaft. Therefore, our first consideration naturally reverts to the beginning of all this action—to the shafts upon which the gears are dependent.

DRAIN

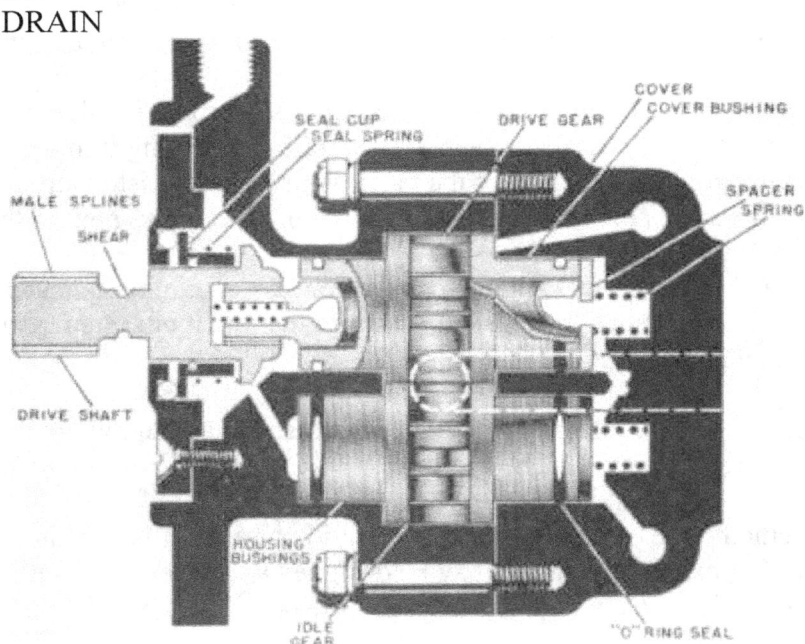

Figure 23.—Schematic diagram of a pressure-loaded gear-type pump.

The ends of both the drive shaft and the idle shaft ride in bronze bushings, which are enclosed in an aluminum-alloy housing. The adapter section of the housing holds the drive-shaft couplings and a drive-shaft positive-type seal.

A shear-pin or shear-section that will break under excessive loads is incorporated in each drive shaft. This

safeguard is designed to protect the pump in case of overload by freeing the coupling from the driving mechanism.

Four check valves are fitted into the housing cover. Two of these valves are connected by communication passageways to the back of the cover bushing-flange to provide loading pressure for the pumps. Such pressure forces the bushings against the gear faces, thus maintaining a minimum clearance between bushings and gears when the pump is in operation. Communication passages and hollow gear shafts act as cooling and lubricating agencies as the fluid, circulated by housing relief-valve pressure, passes through them. These relief valves, set at 15 p. s. i., unload to the inlet section of the pump. This unit is called a pressure-loaded gear pump.

Some gear-pump models use two check valves and two plugs in the cover. Because of this, when changing the drive-shaft direction of rotation, it is also necessary to change the check valves and plugs to alternate bosses. Supply and pressure ports also are changed to opposite positions.

Excessive wear, indicated when the pump fails to develop sufficient pressure, is remedied by removing as many 0.001-inch shims from beneath the cover as are required to adjust these clearances.

When changing the direction of rotation of pressure-loaded pumps having four check valves, it is merely necessary to turn the cover plate 180 degrees. Complete directions for this maneuver are inscribed on a plate attached to the pump. The plate indicating direction of rotation must also be changed when rotation of the pump is reversed.

Gear-Pump Assembly

When disassembling gear pumps, it is advisable to follow instructions contained in applicable maintenance-manuals. However, a few common-sense rules may prove of some help.

The housing and cover should be carefully marked

before the cover is removed to make certain of proper reassembling.

Stud nuts holding the cover plate should be loosened a little at a time. When replacing the cover, draw the nuts gradually, using a cross-tightening method to prevent cracking and distortion of the pressure plate. Final tightening of the nuts should be to the torque value specified in the maintenance-manual.

Remove the cover carefully to prevent loss of parts and to guard against stripping stud bolts.

After thorough cleansing, inspecting, and pre-oiling, all parts should be carefully laid out in order of disassembly.

Care should be taken not to force parts into place. Do NOT DROP BALL CHECKS OR POPPETS ONTO THEIR SEATS WHEN REASSEMBLING HYDRAULIC UNITS.

Gear-Pump Testing Procedure

Testing procedure will vary with different pump models, so overhaul manuals should always be consulted for testing and trouble-shooting instructions. The following example is furnished to familiarize the Aviation Structural Mechanic with general testing procedures.

Hook up the pump and operate it at 1,000 p. s. i. Check the drain ports for leaks. Carefully observe the operation of the pump, making certain that the temperature does not exceed 160° F., and that drain leakage of the adapter is not more than five drops per minute.

It is patently impossible to foresee all conditions which may arise and to prescribe remedial action for all such ills. But a list of the most common ailments—and corrective measures for the alleviation of such troubles—is outlined in table 6.

PISTON PUMPS

The ability to create higher pressure is the advantage of the piston-type power pump over other types of pumps.

The overall efficiency of piston pumps is about 94 percent. It is plainly evident, therefore, that to attain such high degree of operating perfection, these pumps must be very finely machined.

Table 6.— Hydraulic pump troubles and remedies

Pumps made by various manufacturers are different in construction. However, the general principles of operation are similar. The following analysis of the operation of the piston pump shown in figure 24, may help you to understand the operating principles of most of them.

There are seven or nine cylinders spaced equally

Figure 24.—Schematic diagram of a typical piston pump.

around the cylinder-block axis, with the cylinder bores parallel to that axis, as may be seen in the end-view of figure 24. The entire cylinder block assembly (all seven pistons) rotates with the drive shaft, connected by means of the universal link. During each rotation, each individ-ual piston moves up and down in its cylinder, because of the changing distance to the point where it is linked to the drive mechanism.

In figure 24, piston A is shown at what would usually be called the bottom of its stroke. When piston A has rotated to the position held by piston B, it will have moved upward in its cylinder forcing fluid out through the discharge port all the while. During the remainder of the rotation back to its original position, the piston will be traveling downward in its cylinder, drawing in fluid through the inport.

The inport and outport, shown in white in the end-view of figure 24, extend through nearly half of each 180 degrees of operation. Thus each piston passing the inport is drawing in fluid, and each piston passing the outport is forcing out fluid.

This multi-piston pump is the most powerful used in aircraft hydraulic systems, developing up to 3,000 p. s. i. working pressure in some models.

Pump Assembly and Disassembly

Because assembly and disassembly procedures will vary with each pump model, instruction manuals should be studied and understood before such undertaking is begun.

After a unit has been disassembled, a thorough inspection of all parts should be made, and damaged or worn parts replaced. Damaged lapped surfaces, such as the valving plate to the top of the rotating cylinder-barrel, should be relapped or replaced. If relapping will suffice, it is imperative that a perfect operation be performed on the valving plate where it contacts the lapped surface at the top of the rotating cylinder-barrel. This process must be performed with a special lapping block.

Pistons are lapped individually to each cylinder in the cylinder-barrel, and attention is directed to the fact that pistons must not be handled more than is absolutely necessary. Hand moisture will create oxidation which is harmful to lapped surfaces.

All parts must be carefully cleaned before assembly, and your task will be lightened if the parts are arranged in order of assembly.

When the assembly is completed, the pump must be primed through the housing drain. Make certain that the unit is plugged after priming.

Following complete assembly, the piston pump must be bench-tested before reinstallation. The test stand should have its own drive-shaft, incorporating a shear section to protect the pump, even though the pump drive-shaft also has a shear section.

Run the pump for 15 minutes at not more than 100 p. s. i. at 1,000 r. p. m.

Proof-testing is accomplished by plugging the inlet and outlet ports and applying 50 p. s.

i. to the housing drain port. No leakage should show at gaskets, seals, or expansion plugs.

For performance tests, operate the pump at 3,600 r. p. m. under 1,000 p. s. i. outlet pressure and 12 inches of mercury (inlet suction) for 15 minutes. During this period, a low-pressure gauge connected to the housing drain should indicate pressure not lower than 2 p. s. i. and not more than 15 p. s. i.

Check the reduction of delivery in gallons-per-minute from idling pressure to 1,000 p. s. i. This decrease should not exceed three percent. If on checking the piston pump, after reducing pressure to zero excessive resistance to rotation is noticed, the pump should be replaced.

Direction of rotation should correspond to the arrow on the housing-cover plate, but may be changed by turning the housing cover 180 degrees.

The foot valve in the pump is nothing more than a relief valve built into the cylinder bearing-pin which maintains 2 to 15 p. s. i. housing pressure for lubrication purposes.

HYDRAULIC MOTORS

The purpose of the hydraulic motor is to convert hydraulic energy into rotary mechanical energy. Hydraulic motors are used mainly to drive hydraulic turrets and, in some instances, to operate wing flaps.

The seven piston type is the motor most commonly used in aircraft. In general design and construction, it is identical with the seven-piston pump.

Motors may be operated in either direction of rotation and may be instantly reversed without damage if protected by a relief valve with the proper setting.

The motor is controlled by a selector valve, and its speed may be controlled by adjustable restrictors. The motor does not contain an adapter drain and has no foot valve to control housing pressure.

The housing drain port and the reservoir are connected by a direct line.

QUIZ

1. Why are single-action pumps unsatisfactory for use in aircraft hydraulic systems?

2. In a double-action hand pump, what does spongy operation indicate?

3. How often are hand pumps removed for overhaul?

4. Name the two principal groups of power pumps in general use in aircraft hydraulic systems.

5. What is the principal difference between constant-displacement and variable-displacement power pumps?

6. What safety device is incorporated in a gear pump to protect the pump from overload?

7. How may excessive wear which causes insufficient pressure in a gear pump be remedied?

8. What special precaution must be observed in the handling of ball checks or poppets when reassembling hydraulic units?

9. Where should you look for testing and trouble-shooting instructions for gear pumps?

10. What is the advantage of piston-type power pumps over other types of pumps?

11. When hydraulic pumps overheat because clearances are too close or because gears are binding what are the proper corrective measures to be taken?

CHAPTER 6
PRESSURE REGULATORS REGULATOR PRINCIPLES

Regulators are devices designed to control the amount of pressure built up in hydraulic systems and to unload or relieve the hydraulic pump when desired pressure is reached. This action of pressure regulators keeps the power pump from working against a load when the hydraulic system is not functioning. During this time the pressure regulator by-passes fluid back to the reservoir.

Exactly the same thing that would happen to a football if air were forced into it without regard for its capacity, would happen in the hydraulic system if the engine pump went on and on building up pressure. Like the football, the hydraulic system would blow up. If you were to inflate a football by mouth, and then attempt to hold the pressure you had built up by continued force of breath, you would be doing the job the hard way. Therefore, you solve the problem of keeping the ball at full pressure simply by pinching the tube with your fingers.

The problem of avoiding ruptured lines and other damage caused by excessive pressure, and at the same

time maintaining sufficient pressure in the lines to operate hydraulic units, is successfully met by the use of regulators.

REGULATOR FUNCTION

A typical pressure regulator shown in both the cut-in and cut-out position is illustrated in figure 25. In the CUT-IN POSITION, the reaction of the unit being actuated is the resistance which creates pressure. This pressure is pushing against the under side of the unloader piston, but is opposed by regulating-spring tension. Fluid flow from the engine pump to the unit being actuated is keeping the system check unseated.

The drawing of the cut-out position illustrates the position of the regulator when the actuating cylinder has reached the full extent of its travel. The back-pressure now has sufficiently increased to overcome the regulating-spring tension. The pilot pin unseats the unloader check which allows free fluid flow from the pump all the way back to the reservoir.

Several manufacturers have designed valves to serve the two fundamental requirements of regulators, but there are basically only two principal types—the automatic type and the semi-automatic type.

AUTOMATIC TYPE REGULATOR

Let us consider first the automatic type regulator from the viewpoint of purpose and location. Figure 25 illustrates a simple regulator supplying required pressure to a landing-gear system. Study this drawing with care, for we will spend some time learning just how this device works.

In the cut-in position shown in figure 25, trace the flow of fluid from the engine pump, through the regulator and the selector valves, and into the actuating cylinder which raises the landing gear. Notice that when the piston in the actuating cylinder is in motion, fluid enters one end of the cylinder and is displaced by piston

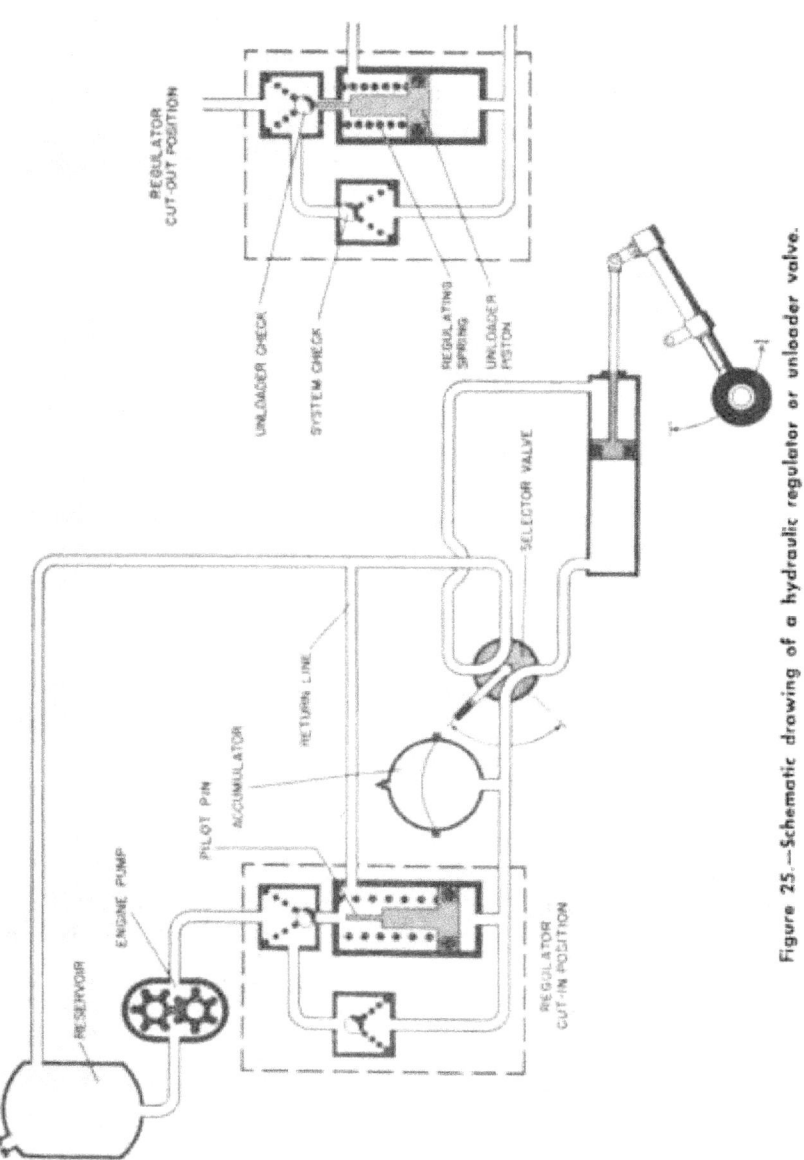

Figure 25.—Schematic drawing of a hydraulic regulator or unloader valve.

movement from the other end of the same actuating cylinder.

Now trace the flow back through the return line and through the selector valve until the fluid finally reaches the reservoir.

Notice the schematic construction of the selector valves. If the selector handle were turned in the direction indicated by the arrow, the fluid flow would be reversed and would enter in and out the actuating cylinder in the opposite direction, thus lowering the landing gear.

The cut-in position of this illustration shows the landing gear being retracted.

Now let us see just how the regulator holds the pressure under control. As the landing gear is raised, pressure in the system is built up all the way back to the engine pump. This means that the pressure also is accumulating on the bottom of the unloader piston. The regulating spring, however, holds the piston down against the pressure.

When the landing gear reaches the end of its travel and can go no farther, what happens? We now have reached the stage where increased resistance has created increased pressure.

In a situation such as this, something has to give and something does give. That something is the regulating spring. At a predetermined setting of the regulating-spring tension, the fluid under pressure forces the piston pilot-pin sufficiently high to unseat the unloader ball. When this action occurs, flow from the engine pump returns to the reservoir, relieving all back pressure on the pump. In other words, the pump has now been unloaded.

System pressure cannot escape because the system check valve is seated. As long as the system pressure is up, such pressure will hold the regulating-spring compressed, thus directing pump flow, under no pressure, to the reservoir. This is called the cut-in position, and is illustrated in figure 25.

If leakage exists in a system after a regulator had

been actuated, pressure beneath the unloader piston would drop. The regulating-spring tension would then push the pilot-pin to one side and allow the unloader ball to seat. Pump flow then would unseat the system check, and pressure would once again build up in the system.

You can see that if the regulating-spring tension is adjusted so as to put the regulator in the cut-out position at a predetermined pressure, a certain amount of pressure also must exist at which that spring tension will allow the regulator to return to the cut-in position. Spring tension at which the regulator must cut in must create sufficient resistance to provide the minimum pressure required to begin unit actuation.

The number of pounds of pressure between "cut-in" and "cut-out" pressure is called "pressure differential," and usually is not adjustable.

When the cut-out pressure is raised or lowered, the cut-in pressure increases or decreases proportionately, thus maintaining the same percentage of differential. The lower pressure is called the "cut-in" pressure, and the higher is termed the "cut-out" or "system" pressure. The system pressure setting will always meet the maximum demands of the system.

One complete operation of the regulator from cut-in to cut-out is called a cycle.

COMPARISON BETWEEN OPEN-CENTER AND CLOSED-CENTER SYSTEMS

We have seen that the pressure regulator maintains a constant high pressure in the system. To understand more clearly the pressure regulator (or closed-center) system, let us compare it with the open-center type.

When we speak of a closed-center system, we refer to the fact that fluid under system pressure is trapped in the system when the selector valves are set in a neutral position. That is to say, no fluid passes through the selector valve.

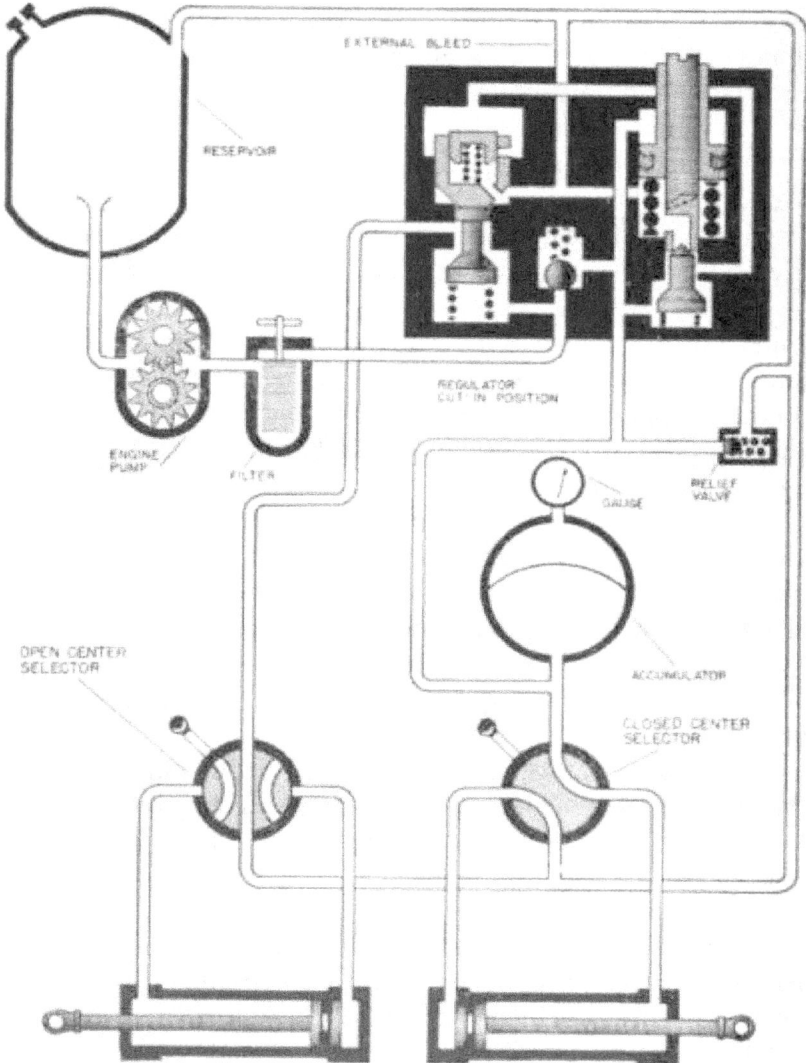

ACTUATING CrLINOCRS

Figure 26.—Comparison between closed-center and open-center systems, showing regulator and selector valve installation and function.

In an open-center system, hydraulic fluid flows from the reservoir, through the pump, through the selector valves when those units are set in a neutral position, and back to the reservoir. This system requires no regulator. In the following chapter on Selector Valves, the function of such valves, as mentioned above, will be fully discussed.

No pressure exists in the open-center system—except that due to fluid friction—when no units are in operation. As may be seen in figure 26, the selector valves are the units which relieve the pump when actuating cylinders reach the end of their piston strokes.

The open-center system consists primarily of a reservoir, an engine pump, a relief valve, a pressure gage, as many open-center selector valves hooked up in a series as are needed for operation of the units, and the actuating cylinders.

As figure 26 demonstrates, external bleed is provided to furnish smoother operation and to allow trapped fluid to escape when the regulator cuts-in while maintaining back-pressure in the return line. The regulator in the cut-out position of this illustration shows the pump flow being directed through the return line and through the open-center selector valves to the reservoir. When the open-center selector valve is turned, back-pressure will not affect the regulator, and therefore will not affect pressure in the closed-center system. When the regulator cuts-in, the return closes, directing pump flow to the closed-center system.

The schematic drawing in figure 26 discloses that a typical pressure regulator system includes the following units: an engine-driven pump for operation of the landing gear, wing flaps, cowl flaps, bomb-bay doors, and brakes; a selector valve to direct fluid from the pump to the system; a pressure regulator to unload the pump and keep the pressure within certain predetermined values; a system relief valve to prevent the development of excessive pressure if the pressure regula-

tor fails to function; and a pressure gage to indicate the amount of pressure in the system.

The general procedure for the operation of mechanism in the pressure-regulator system is to move the selector valve to the desired position, which will cause the pressure regulator to open automatically and relieve the pump when the actuating cylinder has reached the end of its stroke and pressure has built up in the accumulator.

Now that we have observed the operation of typical regulators, we are ready to take up a discussion of several specific types of this unit. This will furnish the Aviation Structural Mechanic with information pertaining to adjustment of these units. The Electrol, Vick-ers, and Bendix regulators described are standard units in Naval aircraft hydraulic systems.

Electrol Regulator

The Electrol regulator has three ports; one system port, one engine-pump port, and one return port. This type regulator is shown in detail in figure 27.

A study of figure 27 will disclose that when the Electrol regulator is in the cut-in position, ball (3) is seated on the small-area hole. No pressure exists in chamber (5) to cause piston (4) to unseat ball (1). Therefore, fluid flow is from the pump port through check valve (2) and into the system.

When the cut-out position is reached, the pressure on ball (3) is sufficient to overcome the force exerted by spring (6). Ball (3) transfers from the small-area seat to the opposite seat, which has greater area. When ball (3) transfers, system pressure is permitted to act on piston (4), which unseats ball (1), allowing fluid flow to be directed to the return port.

The system pressure acts on check valve (2) seating it to prevent loss of system pressure back through the regulator. When the system pressure drops to the cut-in setting, the tension of spring (6) overcomes the

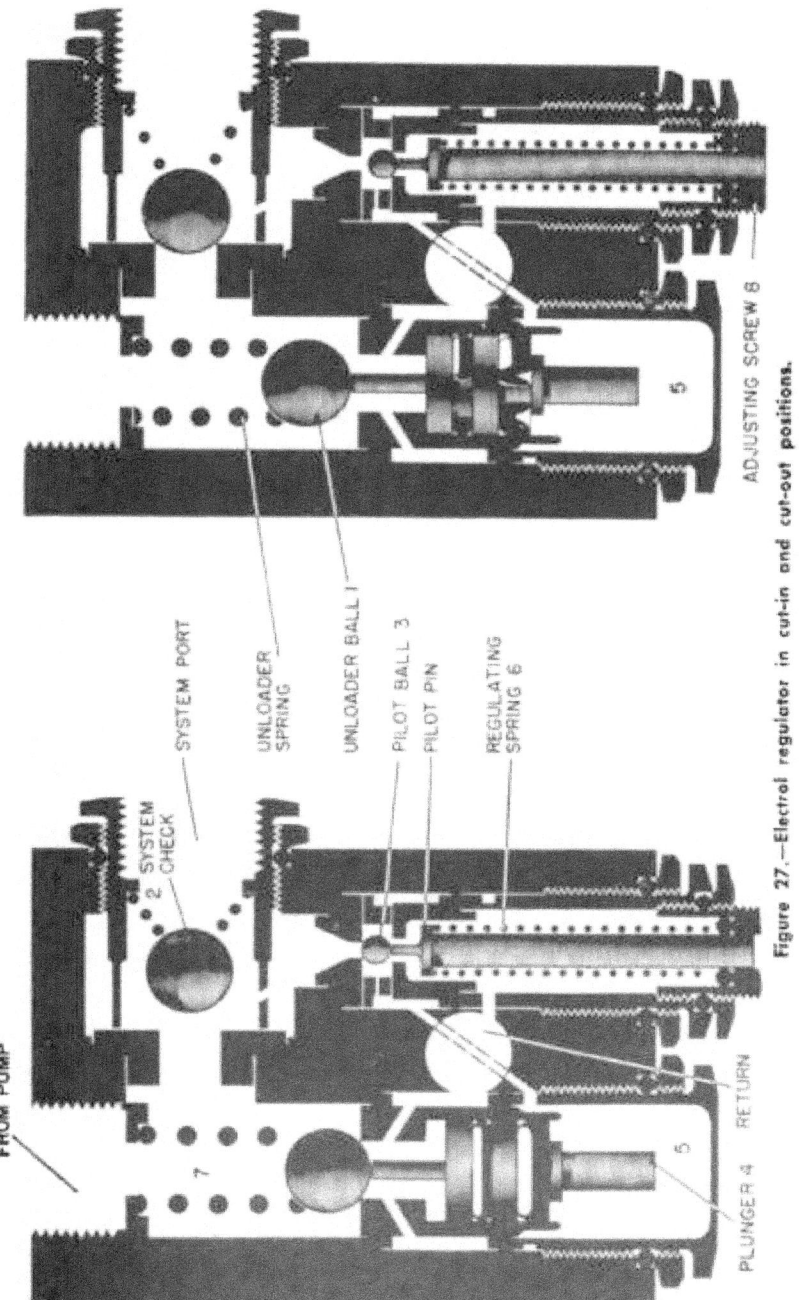

Figure 27.—Electrol regulator in cut-in and cut-out positions.

pressure on ball (3), which transfers to the small-area hole. This transfer allows the pressure trapped in chamber (5) (which lies behind piston (4)) to escape through chamber (7) into the return port. Ball (1) then seats and closes off the return flow of fluid. The flow is then directed through check valve (2) into the system.

The only practical adjustment of the Electrol regulator is that which can be made by tightening or loosening nut (8). This adjustment will raise or lower the cutout setting which in turn raises or lowers the range. No pressure differential adjustment is possible. The area of the seat for ball (4) determines the differential between the cut-in and cut-out pressures.

Vickers Pressure Regulator

The Vickers pressure regulator comes in two basic designs—one with an internal drain, and one having an external drain. Since the type containing the external drain is currently used in naval aviation, that model will receive our attention.

When the Vickers pressure regulator is in the cut-in position, as shown in figure 28, the fluid flow is from the pump port through check valve (2) into the system. When the system pressure reaches the cut-out setting, the pressure acting upon plunger (5) overcomes the tension on spring (6), moving pilot spool (1) upward.

Pilot spool (1) then directs fluid under pressure to the left end of directional spool (4), fluid moving this spool to the right. At the same time, the movement of spool. (1) provides an escape for fluid from spool (4) into line (8), which forces unloading spool (3) from its seat. This permits a free flow of fluid between the engine pump and the return port.

Check valve (2) then closes to prevent the system pressure from escaping back through the regulator to the reservoir. When the system pressure drops to the cut-in setting, spring tension (6) forces pilot spool (1)

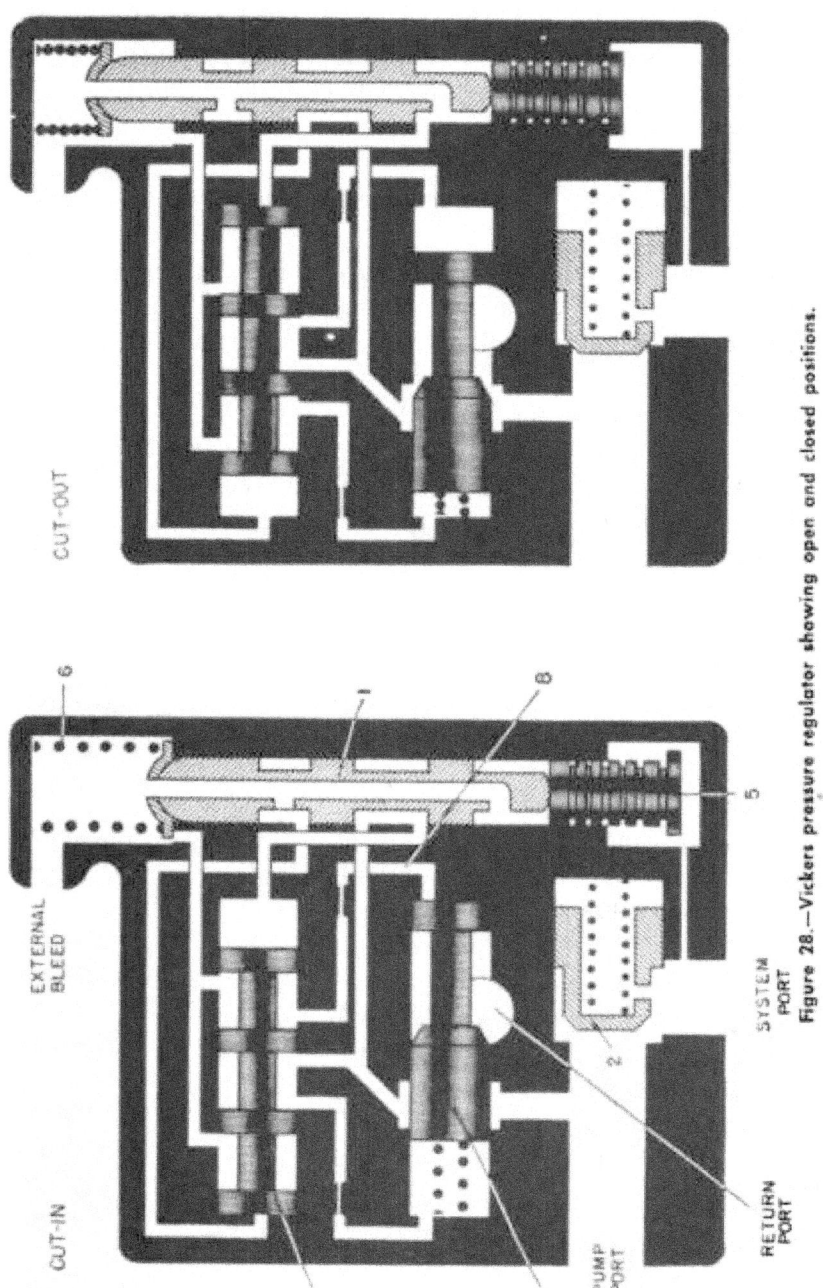

Figure 28.—Vickers pressure regulator showing open and closed positions.

down. This causes the reverse action to occur and the regulator again cuts in.

The Vickers pressure regulator has no external adjustment. To raise or lower the range, shims must be added or removed from beneath the pilot spring valve (6). The pilot spring valve is made accessible by removing the pilot-spring cap.

Bendix Regulator

The Bendix regulator is a high-pressure unit used on 3,000 p. s. i. systems. Its function is to maintain adequate pressure in the hydraulic system by cutting-in (loading) when pressure falls below 2,700 p. s. i. and continuing to operate until pressure reaches 3,000 p. s. L, when it cuts out (unloads). When not operating, the regulator bypasses fluid directly from the pumps back to the reservoir, thus preventing overheating of the fluid and possible damage to the system.

Adjustment of the Bendix pressure regulator may be accomplished as follows: Connect the regulator with a test stand and adjust the pilot valve cylinder so that the main valve plunger will open when 3,000 p. s. i. pressure is applied to the pilot valve pressure port and will close when pressure is reduced to 2,700 p. s. i.

TESTING PROCEDURES

Testing procedures for the various types of regulators we have discussed are outlined in the following paragraphs.

In the Electrol regulator, shown in figure 27, testing procedure is effected as follows:

Apply system pressure at the system port, and check the return and pump ports for leakage. No external leaks should be evident. Plug the system port and apply 90 percent of the system pressure at the pump port. Check for leaks at the return port.

In the Vickers pressure regulator, illustrated in figure

Figure 29.—Bendix pressure regulator.

28, charge to within 50 p. s. i. of the unloading pressure, then remove the engine-pump pressure line and the return line. No more than 10 drops per minute leakage from return and external bleed ports is allowed, and leakage from the pump pressure port should not exceed three drops per minute.

Apply one and one-half times the system pressure and check for external leakage.

SEMI-AUTOMATIC TYPE REGULATOR

Figure 30 shows the SNJ power-control valve which is a manually-controlled regulator. This drawing illustrates the valve in the disengaged position in which the engine-pump flow is being directed through the return to the reservoir.

Figure 30 demonstrates that by engaging the power-control handle, the plunger will be pushed over the metering pin, compressing the return spring and restricting the return flow. Pump flow entering the pressure port is then directed through the hand-pump check valve, the landing-check gear, and the flap check valve.

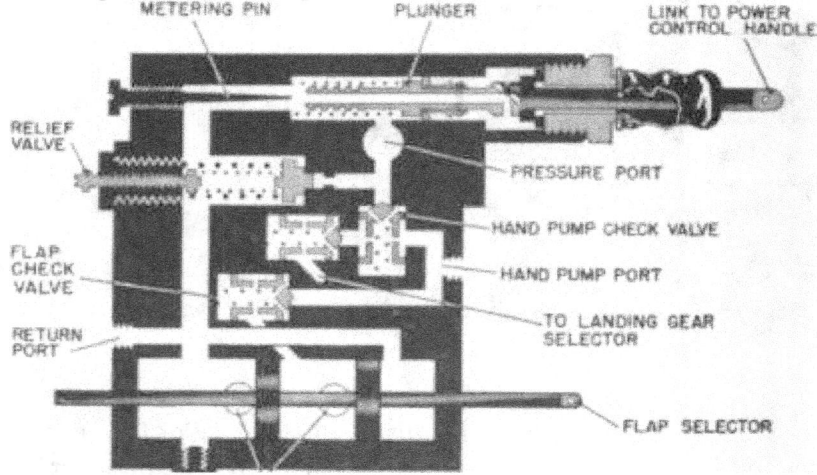

^ m CYLINDER PORTS

Figure 30.—SNJ power-tontrol valve.

During this operation, the landing-gear selector valve, not shown in the illustration, is in the down position. Fluid passing through the flap check valve is directed through the flap selector valve to and from the flap actuating cylinders.

When the pressure rises above 1,000 p. s. i. during flap operation, the relief valve will open to relieve pressure to the reservoir.

Remember that when a unit is being actuated, resistance in the form of pressure goes all the way back to the engine pump.

Look again at figure 30. Notice that the heavy spring around the plunger forces the power-control handle toward the disengage position. This action is slowed by fluid under pressure, metering through the hollow plunger past the metering pin.

TESTING AND ADJUSTMENT

Some difficulty may be experienced with sticking power-control handle and linkage adjustment. Disassembly and cleaning usually will remedy this condition.

From one to two minutes are required for adjustment of the metering valve. Different sizes of metering pins will alter the time lag. A larger pin will cause greater restriction, thereby increasing the time lag, while a smaller pin will decrease restriction and consequently lower the time lag.

It is important to remember that cold weather and cold fluid may increase the time lag. Always allow fluid plenty of time to warm up. Variance in linkage also may create time lag differences.

QUIZ

1. What devices are designed to control the amount of pressure built up in hydraulic systems and to unload or relieve the hydraulic pump when desired pressure is reached?

2. What two principle types of valves are used in regulators?

3. What type of trouble would you expect if in a landing gear hydraulic system, after the regulator has been actuated, pressure beneath the unloader piston drops; the regulating spring tension pushes the pilot-pin to one side and allows the un-loader ball to seat; pump flow unseats the system check; and pressure once again builds up in the system?

4. What term is used to describe the number of pounds of pressure between cut-in and cut-out pressure?

5. What is a cycle of operation of the regulator?

6. What is the purpose of a system relief valve in a typical pressure regulator system?

7. Name the two basic designs of the Vickers pressure regulator.

8. Between what pressures is the Bendix high-pressure regulator designed to operate?

i). In the Vickers pressure regulator, what is the maximum leakage allowed from return and external bleed ports?

10. Is the SNJ power-control valve a manually or an automatically controlled regulator?

CHAPTER 7

SELECTOR VALVES PURPOSE AND CLASSIFICATION

The purpose of a selector valve is to control the direction of the operation of a mechanism by directing fluid under pressure to the desired end of the actuating cylinder while simultaneously directing fluid from the opposite end of the actuating cylinder to the reservoir. Hydraulic fluid flowing through the lines from the pump is routed into these valves which control the direction of its movement as it travels on its way to actuate various mechanisms.

Selector valves may be classified into two groups—

TWO-WAY SELECTOR VALVES and FOUR-WAY SELECTOR VALVES.

The two-way valve is used in conjunction with an actuating cylinder which has a spring to move the piston in one direction and which uses hydraulic fluid to move the piston in the opposite direction. The four-way valve is used with an actuating cylinder requiring hydraulic power to move in both directions. This latter unit is that most widely used.

The open-center selector valve is used in an open-center hydraulic system, which we will consider in a later chapter. All open-center selector valves in an open-center system are connected to common pressure supply and return lines. When the valve is in neutral, the pressure port is open to the return port. When the valve is in the operating position, fluid flows to the actuating cylinder, and then returns to the valve and flows through the open-center return line to the reservoir.

The selector valve used in a direct pressure system is a simple four-port valve that directs fluid from a pressure supply line into the actuating cylinders and discharges returning fluid into a reservoir return line.

In a controlled pressure valve, the output pressure can be controlled within set limits by the operator through manipulation of the valve controls. Such valves must be used in conjunction with an accumulator that maintains a constant pressure at the valve inlet. The power-brake control valve is an example of the application of this principle.

Selector valves are designed to be easily engaged and to allow the passage of the fluid without a pressure drop. Some valves use a portion of the fluid pressure to assist the pilot or copilot in their engagement. These valves are known as balanced valves.

Selector valves may be located in either the cockpit where they are engaged directly, or they may be located in some other part of the aircraft where they are engaged by remote control.

Selector valves are sometimes called control valves. It is true that selector valves may be placed in this classification, but it must be understood that all control valves are not selector valves. A selector valve, in the strict sense of the term, is one that is engaged at the will of the pilot or copilot for the purpose of directing the fluid to the desired unit. In other words, a selector valve is

one which the pilot uses to select the route in which he wants the fluid to go.

ROTOR-TYPE SELECTOR VALVE

The rotor-type selector valve, a unit very similar to the type of valve used in aircraft fuel systems, has four ports—one pressure port, one return port, and two cylinder ports. These ports are located 90 degrees apart around the circumference of the valve housing. The inport, leading from the pressure manifold, and the out-port, connected to the pressure manifold, are always opposite, or 180 degrees apart. The cylinder ports, also separated by 180 degrees, connect to the opposite ends of an actuating cylinder.

The rotor contains a pair of convex channels, designed to connect with two adjacent ports. These channels direct fluid from the pump to the actuating cylinder. This control of fluid is effected by movement of the selector handle to which the rotary plug is attached.

Return fluid from the opposite end of the actuating cylinder is directed from the cylinder port and thence back to the reservoir. When the valve is in neutral— that is, when the rotor is turned so that the channels

PORT (B)
PORT (A)
TO PRESSURE

PORT (D) TO RETURN

PORT (C)

§|||||f FLUID UNDER ™™ PRESSURE

n FLUID UNDER ^"ATMOSPHERIC PRESSURE

Figure 31.—Typical roter-type selector valve.

do not align—the flow is stopped, thereby designating this a closed-center type selector valve. This feature is valuable in the case of wing and cowl flaps when only partial actuation of the flap is desired.

To make certain that we understand the principle of selector valve function, let us observe one as it actually performs. We have seen that each selector valve contains four ports—a pressure and a return port 180 degrees apart, and one port for each actuating cylinder line. We also have seen that the selector-valve handle turns the rotor in and out of alignment with these four ports.

We will consider a selector valve directing fluid flow into an actuating cylinder which operates a wing flap. Let's turn the selector-handle to the down position and see what results.

The movement of the handle turns the rotor into aline-ment with the pump pressure port and the actuating cylinder line which extends the flaps. We see that the piston in the actuating cylinder cannot move if fluid in the other end of the cylinder is trapped. The fluid can, however, pass back through the selector valve up-port, into the return port, and back to the reservoir.

To put the flap up, we turn the selector-valve handle to the flaps UP position. This action reverses the flow of fluid to and from the flap-actuating cylinder, forcing the flaps upward.

POPPET-TYPE SELECTOR VALVE

Poppet-type selector valves are classified as in-line and coaxial in design. A comparison between the types is clearly drawn in figure 32. This classification may, however, be broken down into three additional groups known as balanced cone, unbalanced cone, and ball types. Balanced and unbalanced-type selector valves will be discussed in detail later in this chapter.

The essential purpose, use, and operation of the ball-type selector valve is identical with that of poppet-type valves. The main difference is implied in their names—

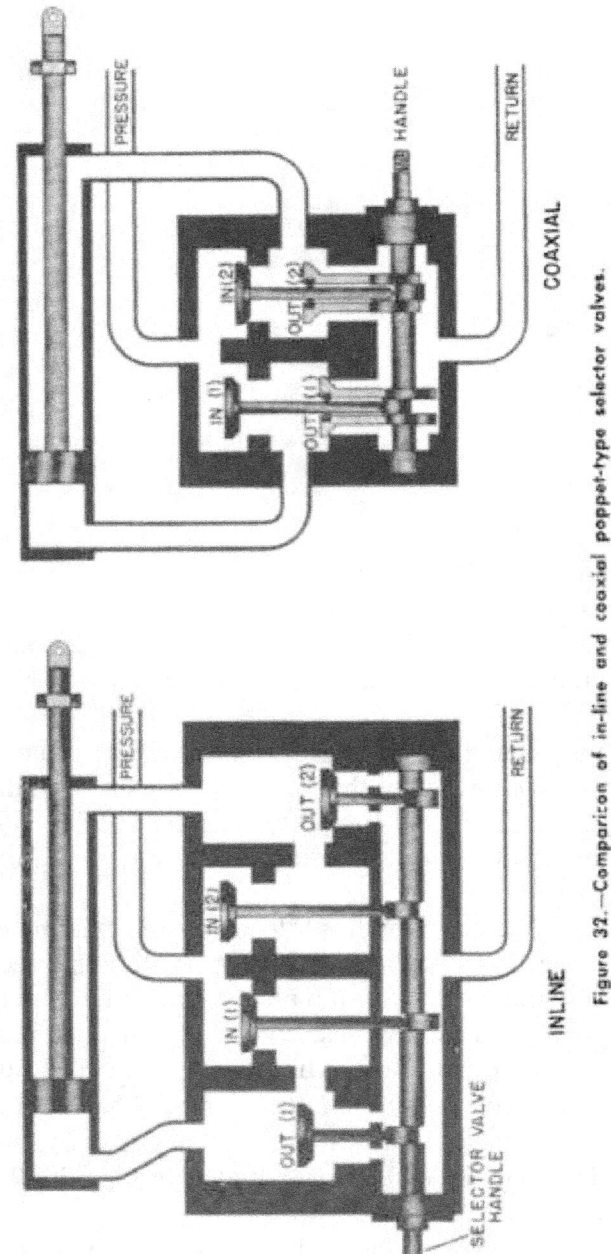

Figure 32.—Comparison of in-line and coaxial poppet-type selector valves.

one controls fluid flow by the use of poppets while the other employs a ball-shaped valve for this purpose.

PISTON-TYPE SELECTOR VALVE

Flow in the piston selector valve is directed by the sliding action of a piston which is actually a hollow spool-shaped plunger within the valve housing. This piston contains the usual four ports found in all selector valves. Drilled passages in the piston connect the hollow section to each end of the housing.

OPEN-CENTER SELECTOR VALVE

Prominent among piston-type selector valves is the "Aircraft Accessories" open-center piston-type selector valve. The salient feature of this valve is that it returns to neutral automatically when the actuating cylinders have reached the end of their strokes.

The open-center valve serves the dual purpose of directing fluid under pressure to one end of an actuating cylinder while simultaneously directing oil from the other end of the cylinder to the return line. When this unit is in the neutral position, fluid flow is directed through the unit to the reservoir.

Figure 33 illustrates an open-center selector valve in selection position, and in figure 34, the same valve is shown in neutral position. The indexing mechanism of the open-center selector valve is a spring-loaded cam arrangement bolted to one end of the selector housing. This mechanism serves the three important functions of holding the housing in neutral position; holding the piston in either working position desired; and returning the piston to a positive neutral.

The pressure scoop in the piston directs the flow of fluid through the housing, through the cylinder port, and into the actuating cylinders. Return flow enters through the opposite cylinder port to the return scoop in the piston which directs flow to the reservoir. At the end of the

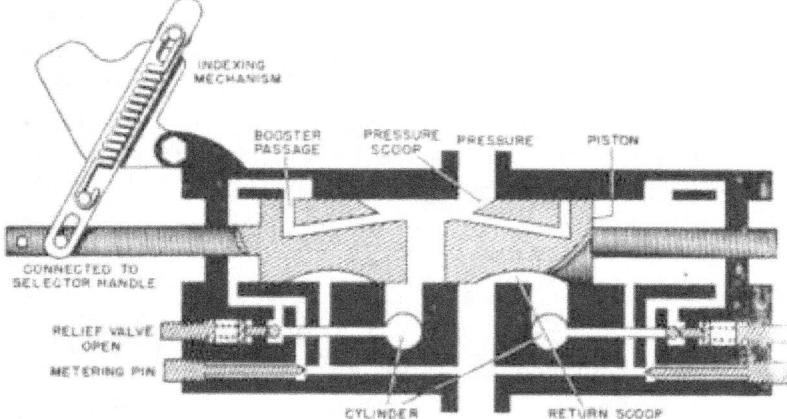

PORTS RETURN

Figure 33.—Open-center selector valve in selection position.

actuating cylinder travel, the sudden increase of pressure opens the relief valve and exerts pressure on the end of the piston which starts it toward neutral; after the piston moves a short distance, it uncovers the booster passage which gives greater impetus to returning the valve to neutral. In this position, flow of fluid goes directly to the return line.

INDEXING MECHANISM

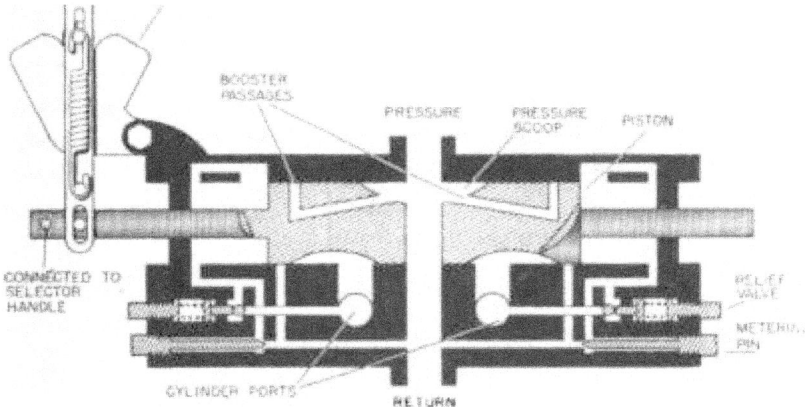

Figure 34.—Open-center selector valve in neutral position.

Relief valves are most important features of this type of selector valve. Two adjustable relief valves are located in the housing of each selector valve and serve the threefold purpose of relieving thermal expansion from cylinder lines; insuring full operation of units actuated ; and starting the piston on its way back to normal from a working position. This unit is nothing more or less than a safety valve keeping a watchful eye out for excessive pressures.

The relief valves in the open-center selector unit are set at 150 p. s. i. above unit operating pressure. To illustrate this point, let us say that it would require 850 p. s. i. to raise the landing gear. We therefore would set the UP relief valve to relieve at 1,000 p. s. i. at which point the valve would return to neutral.

The adjustment of relief valves before installation of new units should be performed on a hydraulic test bench according to the following outline.

Place the selector valve in neutral and apply pressure to the cylinder port nearest the relief valve on which work is being performed. Keep an eye on the gage to note the pressure at which the valve is relieving. Turn the adjusting screw IN to increase relief pressure, and OUT to decrease the pressure.

If for any reason it becomes necessary to increase or decrease the setting while the valve is mounted on the plane, place the selector valve in working position and build up pressure with the hand pump. Watch the system pressure-gage for the amount of pressure being built up until the selector valve kicks back to normal, or neutral. The mechanic will thus be enabled to determine the direction in which the relief valve adjusting screw must be turned.

The number of increasing or decreasing turns applied to the screw is a matter of trial and error, since no definite number of turns to achieve any specific pressure can be accurately considered as standard.

The only servicing required by these valves is thorough lubrication of the indexing mechanism upon each inspection to prevent binding.

If the selector valve fails to respond when a working position is indicated, and the indexing mechanism is well oiled and greased, the trouble may lie in a clogged metering-pin orifice. If the selector valve will not return to neutral, the relief-valve setting may be too high, or the metering pin may be too far open.

Metering pins are located on the lower end of the selector-valve housing. (To adjust metering pins, screw the unit in completely and then back off taking one-eighth turns until the proper adjustment has been effected.) The main purpose of the metering pins is to build up sufficient back-pressure behind the piston to start the piston toward neutral.

Now let us review the operation of an open-center selector valve by observing its performance in raising the landing gear. In this operation, our first act is to move the selector handle forward. This causes the piston to move forward in the selector housing, and directs fluid to the UP port of the landing-gear cylinder, raising the gear.

When the landing gear is up and locked, pressure builds up to the relief-valve setting; the relief valve opens, and fluid is directed to the forward end of the piston and also to the metering pin. The metering pin is so slightly open that its action builds up pressure which starts the piston toward neutral. When the piston has moved but a short distance, the booster passages line up with slots in the housing to allow full system pressure to gather behind the piston. This pushes the piston still farther toward neutral.

This is where the indexing mechanism takes over. Its spring-loaded cam snaps the selector-valve piston into positive neutral, holding it there until the selector handle is again moved. We now have a free flow of fluid once more passing through the selector valve.

FLEETWING PRE-SELECTOR

The fleetwing pre-selector is a piston-type selector valve of the closed-center type. Its operation is quite different from that of the open-center unit we have just discussed. The outstanding feature of this selector valve is its ability to prevent greater than 3 degrees of flap-creepage in either direction. It also returns the flaps to the original flap-setting.

The follow-up has five O-ring seals. The two outboard seals are the most important because they prohibit external leakage. The follow-up arm is connected by linkage to the flap so that whenever the flap moves, the follow-up is also put in motion.

The selector piston controls the desired degree of flap movement, its range extending from zero to 50 degrees. The selector-valve housing is so designed that the return port is located next to the selector handle, while the pressure port is situated between the cylinder, or working,

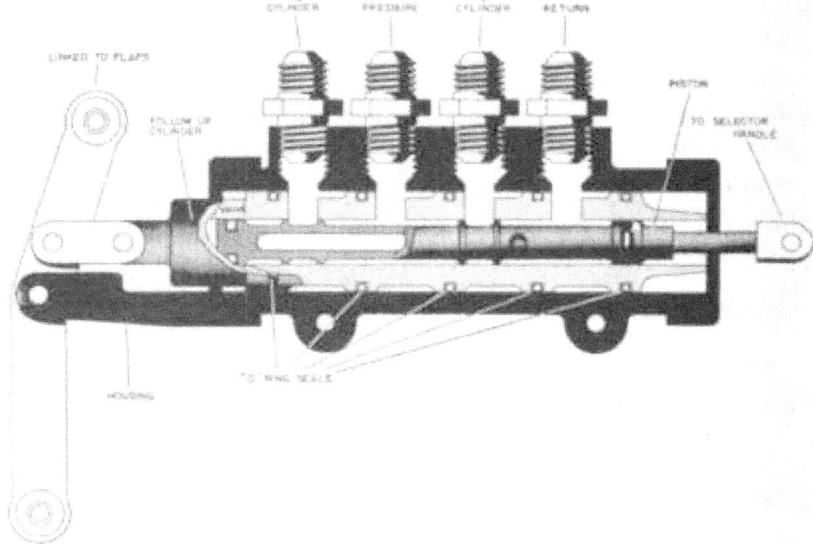

Figure 35.^Fleetwinjj pre-selector valve.

ports. The down-port is that aperture nearest the follow-up arm, and the up-port lies between the pressure and return ports.

Let's see how this pre-selector valve functions by observing it at work when attached to wing flaps. When the selector handle is pushed inward, fluid from the pressure port is directed into the desired cylinder port. Return fluid from the opposite end of the actuating cylinder is then directed to the return port. The outer cylinder is mechanically attached to the flaps which extend it outward and close the port when the flaps have attained their full extent of travel. Once the selector handle is put in selection position, it remains there until manually moved.

As we have noted, the follow-up contains five O-ring seals. When flap creepage occurs, the trouble may lie in these seals, and the following procedure may assist the mechanic in trouble-shooting of this nature.

Number the five O-ring seals in the follow-up from left to right so that numbers one and five are outboard. With the flaps down and the valve in neutral, if number two seal is leaking, the flaps will creep upward. This means that the flaps will not hold the desired positions but will move out of place. We have learned that when the flaps move, the follow-up also moves. This action lines up the pressure port with the down-port and puts the return port in line with the up-port, causing creepage.

A turnbuckle is attached to one end of the linkage between the flap and the follow-up arm. One revolution of this turnbuckle will change the movement of the flaps 10 degrees. A loose bolt in any of the linkage may affect the flap movement as much as 30 degrees. It is therefore important that the linkage be checked at regular intervals.

Fleetwing Test Procedure

Testing procedure for the Fleetwing pre-selector consists of a three-step process.

1. Put the valve in neutral; apply 1,000 p. s. i. pressure to the pressure port, and check all other ports and vents for leaks.

2. Plug cylinder port number one and move the piston inward one-eighth inch. Apply 1,000 p. s. i. pressure to the pressure port and check for leaks at all ports and vents.

3. Plug cylinder ports one and two and move the piston to a completely extended position. Then apply 1,000 p. s. i. pressure to the return port. There should be no external leakage, and internal leakage is limited to 10 drops per minute.

AIREX BALL-TYPE SELECTOR VALVE

The Airex ball-type selector valve belongs to the unbalanced poppet-type category because the balls act as poppets in controlling fluid flow through this unit.

To remember locations and destinations of lines in an unbalanced-type selector valve, bear in mind that the pressure port leads to the top of the pressure poppets; each of the two cylinder (or working) ports lead to the bottom of one pressure poppet and to the top of one pressure poppet, and the return port runs to the bottom of both return poppets.

To clarify the above, let's look at a new Airex-type selector valve with all ports marked, as shown in figure 35. The pressure cylinder and return ports may be identified with the aid of the following experiment.

With the valve in neutral, apply pressure to one of the ports. If no fluid appears from any port, the line is in the pressure port. If fluid emerges from one port, the line is in one of the cylinder ports. And if no fluid comes from two ports, the line is in the return port.

The Airex selector valve is constructed so that the ports are 90 degrees apart. The pressure and return ports are directly opposite, or 180 degrees apart, and the two cylinder ports are similarly located. For the sake of illustration, suppose that each of the ports in this valve

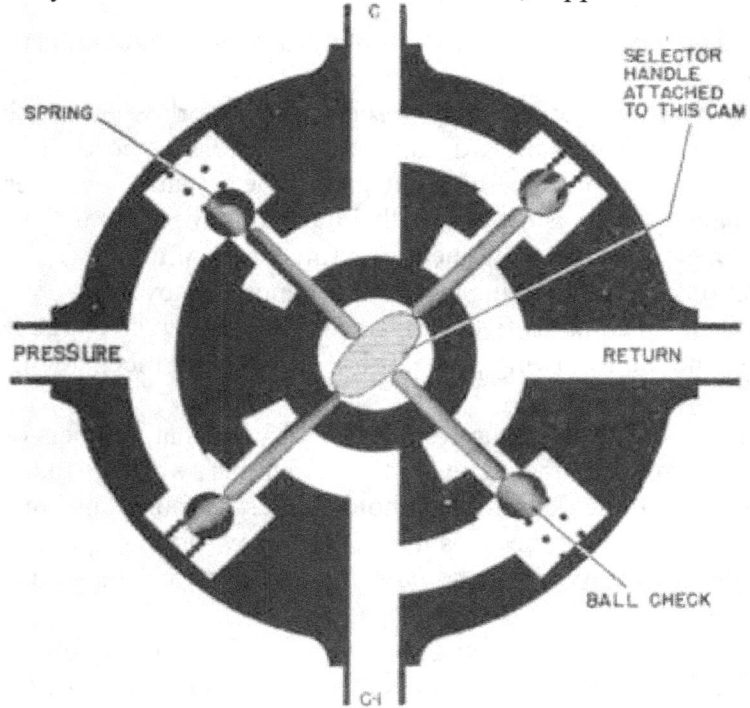

Figure 36 —Airex selector valve.

face a cardinal compass point. The pressure and return ports would be exactly opposite, as east and west, while the two cylinder ports would be to the north and south. If straight lines were to connect each of these opposite ports, the result would be a cross, explaining why the law of position regarding this type of valve is called the "cross" or "X" rule.

The Airex selector valve operates with a cam arrangement in its housing working on pins which make contact with the balls. These, in turn, unseat when the cam depresses the pins, allowing fluid to flow past them. When the valve is in a working position, and one pressure poppet (or ball) is unseated, the opposite return poppet also becomes unseated. The selector handle is moved to neutral by the operator when the actuating cylinder has reached the desired length of travel.

When the pressure in the working lines becomes greater than the system pressure, it will unseat the pressure poppets and relieve through the main system relief valve, thus providing for thermal expansion.

Airex Test Procedure

With all ports uncapped and the selector handle in the neutral, or closed, position, apply pressure of 200 p. s. i. and 2,250 p. s. i. successively at the pressure port. No external leakage at cylinder ports C and C-l should be evident.

With cylinder ports C and C-l capped, turn the handle 22.5 degrees from neutral in either direction and apply the same pressure. There should be no external leakage, or leakage at the return port. Turn the handle 22.5 degrees in the opposite direction and repeat the test.

With all ports except the return port capped, apply the above pressures at return port. No external leakage should be in evidence. Ten drops per hour internal leakage is allowed.

ADEL SELECTOR VALVE

Another unit widely used in Naval aviation is the Adel selector valve which uses unbalanced poppets. This valve may be stacked—that is, it may have more than two cylinder ports but still contain but one pressure port and one return port. This may be accomplished in two ways.

One way is to construct the housing for the unit with an additional number of cylinder ports. The other method is to bolt several units together, as in some planes where two or three valves are stacked with common pressure and return ports.

The Adel selector valve has four cone-shaped poppets —two functioning as pressure, and two as return poppets. The CROSS or X rule as to location applies to this valve as in the ball-poppet type. There is no standard of construction for the valve housing. In fact, several valves

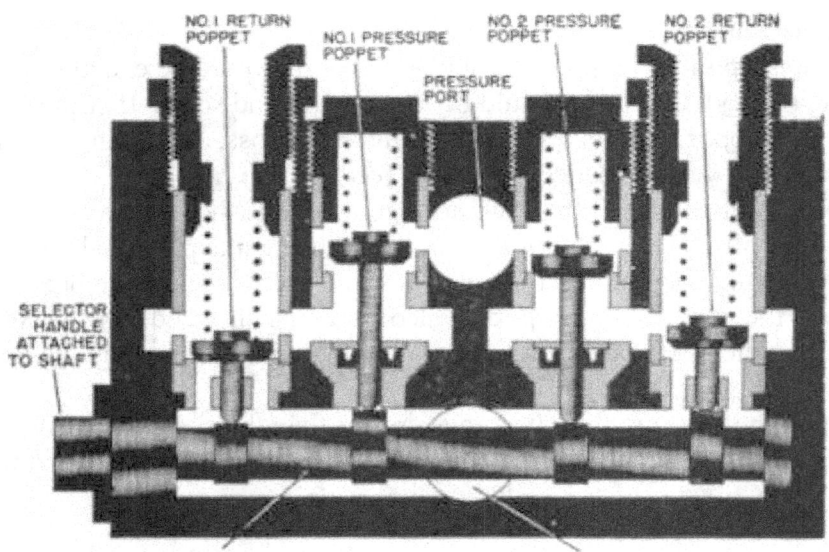

CAMSHAFT RETURN PORT

Figure 37,—Add selector valve.

of this type with varying designs are currently in use.

Basically, however, the Adel selector valve operates similarly to other units of this type. It contains a camshaft with four lobes, each lobe being set to perform the apparently paradoxical function of unseating a poppet and permitting a poppet to remain seated when the camshaft is turned. In this valve, the return poppet opens

TWO DEGREES BEFORE THE OPPOSITE PRESSURE POPPET

OPENS. This prevents line chatter or hammering which may occur in other types of selector valves.

The number one return poppet in the Adel selector valve is located nearest the selector handle. Next to it lies the number one pressure poppet, followed in turn by the number two pressure and the number two return poppets.

Adel Valve Operation

When the selector handle is turned to the right, the number one pressure and number two return poppets

are opened. Remember, the return poppet opens 2 degrees before the pressure poppet responds. Fluid from the return end of the actuating cylinder is now free to return to the reservoir. The flow entering the other end of the actuating cylinder moves the piston downward, forcing the return flow into the valve and back to the return port. When the gear actuated reaches the desired position, the operator puts the valve in neutral.

To reverse this action, turn the selector handle to the left, thus opening the number one return and the number two pressure poppets. This moves the actuating cylinder in the opposite direction.

The Adel selector valve relieves thermal expansion in the same manner as the ball-type unit, by unseating the pressure poppets when pressure in the working lines becomes greater than system pressure. Such relief is effected through the main system relief valve.

ADEL SELECTOR VALVE LEAKAGE TEST

The limit for internal leakage allowance is 10 drops per hour. No external leakage is permitted.

To test this valve for leakage, attach the pressure port to the pressure-line valve on the test stand. Set the handle in the neutral position. Apply 2,500 p. s. i. pressure and check for leakage at poppet caps and at both cylinder ports.

Plug both cylinder ports, set the camshaft in operating position, and apply 2,500 p. s. i. pressure to the pressure port. Check for leakage at the return port.

Set the camshaft in the opposite operating position and, with the same pressure application, check for leakage at the return port.

Plug the pressure port and both cylinder ports. Apply 250 p. s. i. pressure at the return port. Check for leakage around the camshaft. Drop the pressure to 5 p. s. i. and recheck.

ELECTROL SELECTOR VALVE

The Electrol four-way closed-center balanced-type selector valve is unlike any valve we have thus far discussed, in that it has balanced-type poppets.

In the study of selector valves containing unbalanced poppets, we observed that pressure acting on the bottom of these poppets tended to unseat them because of unbalanced working areas. The opposite is the case with balanced poppets. Pressure applied under the poppet acts on the under side of the poppet-head. But—this poppet has an enlarged area on the lower end of the stem which is equal to the area on the under side of the head. Figure 38 illustrates this difference. Pressure exerted upward on this valve is exactly balanced by pressure exerted downward. Because of this balance, pressure has no

EFFECT ON THIS POPPET.

When the Electrol valve is in neutral, no amount of pressure applied to any area can unseat the poppets. Therefore, to create fluid flow through this unit, some method of unseating the poppets is required.

UNBALANCED BALANCED Figure 38.—Comparison of balanced and unbalanced poppets.

Such action is obtained by a camshaft arrangement very similar to that used in unbalanced-type valves, except that the shaft in question has but two lobes and the pressure and return poppets unseat simultaneously. These poppets are so arranged that the two pressure poppets are located directly opposite each other and nearest the selector handle.

The return poppets also are across from each other and are located at the opposite end of the selector valve. The ports are to be found on the top side of the valve, with the pressure port next to the selector handle. The two cylinder ports are on either side the valve, while the return port lies at the far end of the unit.

Electrol Selector Valve Operation

Holding the selector valve with the handle facing you, movement of the handle to the right will open the right pressure poppet and the left return poppet, thus directing fluid flow from the pressure port to the right cylinder port and to the actuating cylinders. When the actuating cylinder reaches the full extent of its travel, the valve remains in a working position.

The poppet slides into a spool containing 16 holes or passageways. Eight of these passageways lead from the pressure and return ports to the top of the poppet. The remaining eight lead from the cylinder port to the inside working area of the poppet. With the selector handle in neutral, 8 to 10 degrees of handle movement should be present.

Electrol Selector Valve Leakage Test

The operating pressure of this valve is 1,500 p. s. i., and the proof pressure is 2,250 p. s. i.

Attach the pressure port to the pressure-line valve on the test stand. Set the camshaft in neutral position, apply 2,250 p. s. i. pressure, and check for leakage at both cylinder ports, pressure caps, and camshaft pressure seals. Leakage from the return port will indicate that the center seal on the camshaft is leaking.

Plug both cylinder ports, set the camshaft in. operating position, and apply 2,250 p. s. i. pressure to the pressure port. Check for leakage at the return port. Set the camshaft in the other operating position and repeat the test.

Plug the pressure port and both cylinder ports. Apply a pressure of 500 p. s. i. to the return port, and check

TO CYLINDER

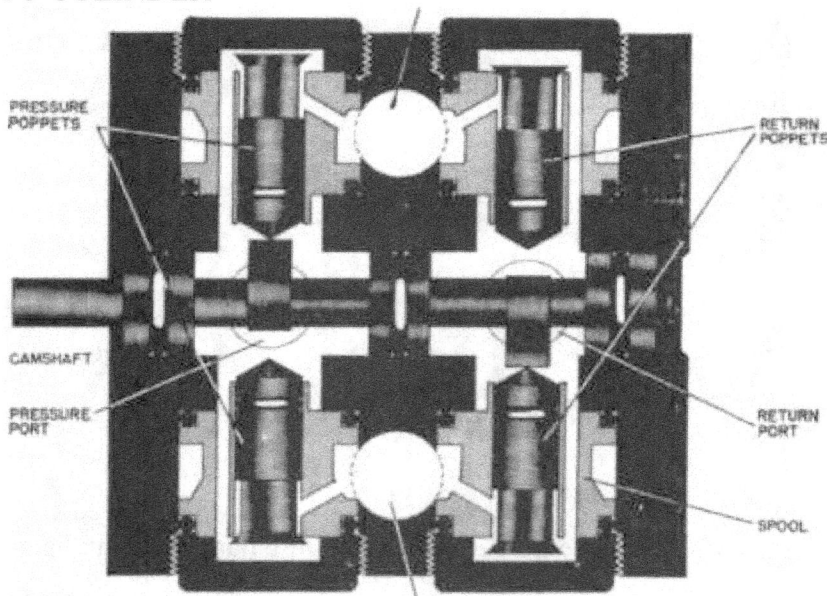

TO CYLINDER

Figure 39.—Electrol selector valve.

for leakage at the return port. Set the camshaft in the other operating position and repeat the test.

Plug the pressure port and both cylinder ports. Apply 500 p. s. i. pressure to the return port, and check for leakage at the return caps and camshaft. Drop the pressure to 5 p. s. i., and repeat the test.

Internal leakage must not exceed 10 drops per hour. No external leakage is permitted.

ELECTROL HAND PUMP SELECTOR VALVE

The purpose of the Electrol hand pump selector valve is to direct the flow of fluid from the hand pump to one unit system at a time in case of emergency or for ground checking.

The construction of this valve is similar to that of the Electrol four-way unit. That is, it uses the same type spool and poppet assembly which may, if necessary, be interchanged.

The Electrol hand pump valve has four working positions, one pressure line, and four working lines. This unit has NO return port. The hand pump valve directs fluid from the hand pump to another selector valve such as the type just discussed.

Using the TBM system as an example, operation of the hand pump selector valve is explained as follows:

The working positions may be numbered from one to

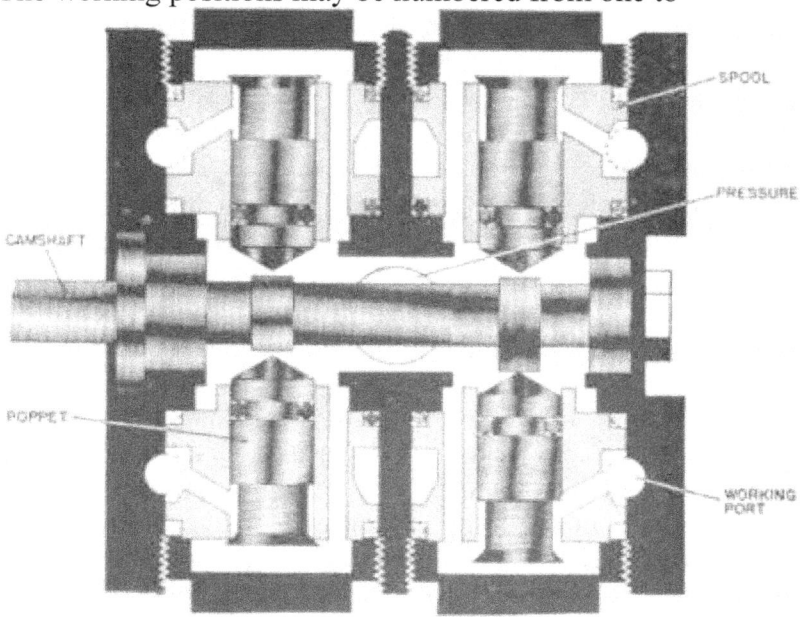

Figure 40.—Electrol hand pump selector valve.

four, and will comprise, respectively, landing gear general (which includes wingfold, cowl flaps, and oil cooler) ; bomb-bay DOORS, and FLAPS.

The Bureau of Aeronautics prescribes that, in normal flight, the selector valve be held in the number two or general, position.

When the hand pump selector valve is held in the number two position, use of the hand pump to create flow will direct fluid to the cowl flap, oil cooler, and wing-fold selector valves. By putting any of these selector valves in working position, the units may be actuated. This also holds true for the working positions of each unit system selector valve.

A simple leakage test for this valve—which may be performed without removing the valve from the plane— may be accomplished in two ways.

With the hand pump selector valve in a working position, kill the system pressure by actuating a unit until the gage registers zero pressure. Place the hand pump selector valve in the neutral position and build up a system pressure of 1,500 p. s. i. If the pressure drops off, the valve is leaking. Constant pressure indicates that no leakage is present.

A similar test may be made with the engine pump. Kill the system pressure at zero reading on the gage. Put the hand pump selector valve in neutral and start the engine. The gage should continue to read zero pressure if the hand pump selector valve is not leaking. If leakage is present, the gage will indicate pressure to a degree and rapidity of rise dependent upon the amount of leakage.

If in either test the valve is found to be leaking, it should be removed from the plane and a more thorough test made on the test stand.

Electrol Hand Pump Selector Valve Test Procedure

The operating pressure of this valve is 1,500 p. s. i., and the proof test pressure is 2,250 p. s. i.

Attach the pressure port (center port) to the test stand pressure line. Set the camshaft in neutral, thus allowing all poppets to be seated. Apply 2,250 p. s. i. pressure and check for leakage at all four ports, caps, and at all pressure seals on the camshaft.

Plug port number one, set the cam at the number one (landing gear) position, and apply pressure. Check for leakage at ports number two, number three, and number four. Repeat this procedure by plugging ports two, three, and four, and placing the cam on working positions two, three, and four. Check for leakage at all ports.

No external leakage is allowed, and internal leakage is restricted to 10 drops per hour.

VALVE-SEAT LAPPING

Before leaving this discussion of selector valves, several pointers are offered here on valve-seat lapping procedures. This operation is as vital a phase of maintenance as is the care of seals and fluids.

Figure 41 illustrates the manner in which the lapping of valve seats may be accomplished. The figures shown in the drawing illustrate examples of this operation before this procedure is undertaken.

An assortment of steel bearings may be welded onto small rods to be used for the preliminary grinding of scats. The bearings may be dipped into a suitable lapping compound and then placed on the seat and moved with a twisting motion applied with very light pressure. This action will enlarge the seat area, making it necessary to grind down the face of the seat in order that the seat area be reduced to its original size. When grinding down the seat-face, a lapped rod of steel may be applied, using a bushing as a guide. Very light pressure should also be applied in this operation.

Various tools may be devised to hold poppet valves while seating. Operations very similar to those used for grinding seats and faces are applied to seating.

Relapping of cylindrical surfaces is not recommended

BEARING ROD LAPPING TOOL

SEAT WIDTH

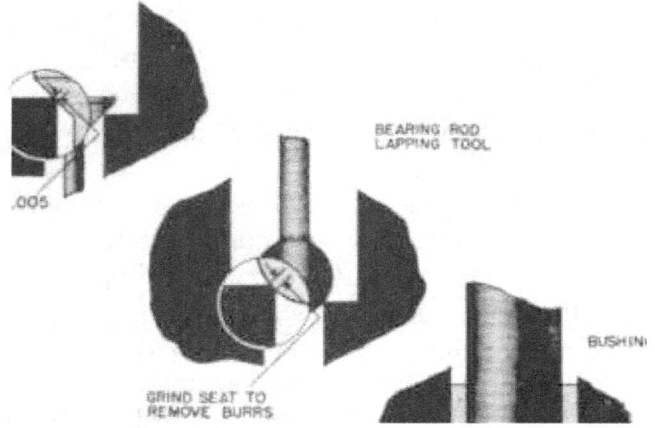

.005

BEARING ROD
LAPPING TOOL

BUSHIN

GRIND SEAT TO
REMOVE BURRS

BUSHING

GRIND SEAT TO REMOVE BURRS

RETURN SEAT TO .005 WfTH FLAT FACED LAPPING ROO

Figure 41 . Valvc-seat lapping procedure.

for inexperienced hands. In cases where such lapping is necessary, the unit or assembly should be replaced.

It is imperative that a unit be perfectly cleaned following a lapping operation. When the work is cleaned, it should be carefully inspected with a magnifying glass for scratches or other imperfections.

When replacing cone-type or ball-type poppet valves,

DO NOT DROP THEM ON THE SEATS.

QUIZ

1. What is the purpose of a selector valve?

2. How does the selector valve accomplish its purpose?

3. Into what two groups may selector valves be classified?

4. The rotor-type selector valve has four ports. Name them.

5. As to design, what two types of poppet-type selector valves are in common use?

6. What device is employed in piston-type selector valves to direct flow?

7. What feature of the "Aircraft Accessories" open-center piston-type selector valve differentiates it from other piston-type selector valves?

8. What functions are accomplished by the relief valves in the open-center selector unit?

9. For how many pounds per square inch above operating pressure are relief valves set?

10. How are metering pins adjusted?

11. What is the outstanding feature of the fleetwing pre-selector?

12. To test the fleetwing pre-selector, how much pressure is recommended?

13. Why does the Airex ball-type selector valve belong to the unbalanced poppet-type category?

14. In the Adel selector valve, what feature prevents line chatter or hammering?

15. What is the operating pressure of the Electrol selector valve?

16. For the TBM system, what position is prescribed by the Bureau of Aeronautics for the selector valve in normal flight?

CHAPTER 8

ACTUATING CYLINDERS PURPOSE

If you are a sports fan, you know that very rarely does the football lineman with the hulking physique also star in tennis. Equally rare is the weight-lifting champion who is also adept in the hundred-yard dash. Nor do wrestlers excel as adagio dancers.

Each athlete, to perform successfully in his particular sport, has had to develop distinct sets of muscles peculiarly adapted to his special activity. And thus it is in hydraulics. The "muscles" of the hydraulic system— the actuating cylinders—have had to be carefully developed for the particular actuating required in various units. Actuating cylinders are available in many shapes and sizes, because their peculiar style must be especially designed to fit the particular job to be done.

It is obvious that a fairly large and powerful actuating cylinder would be required to raise or lower a landing gear, while a smaller unit will suffice for the opening and closing of oil-cooler doors. But the duties assigned hydraulic actuating cylinders in aircraft are consider-

ably more complex than the mere motions of raising, lowering, opening, closing, folding, and spreading.

Let's consider several types of cylinders and study their general characteristics. But let us bear in mind that the purpose and operation of all types of cylinders will remain standard, with differences existing only in construction. Therefore, the units discussed in this chapter are to be considered as representative of all types currently in use.

The purpose of actuating cylinders is to convert fluid energy into mechanical energy. They are used where linear motion is required to move some mechanism.

TYPICAL ACTUATING CYLINDER

Actuating cylinders usually are double-acting, which means that fluid under pressure can be applied to either side of the piston to provide movement in the corresponding direction. Single-acting cylinders are units with a spring return, and are sometimes used to actuate brakes and to charge guns.

Actuating cylinders consist essentially of a cylinder, one or more pistons and piston rods, and the necessary seals. A typical type actuating cylinder, shown in figure 42, illustrates the common construction of all types.

The cylinder in figure 42 operates under the influence of fluid flow in either direction. The cylinder is closed at both ends. Inside is a piston which operates a piston

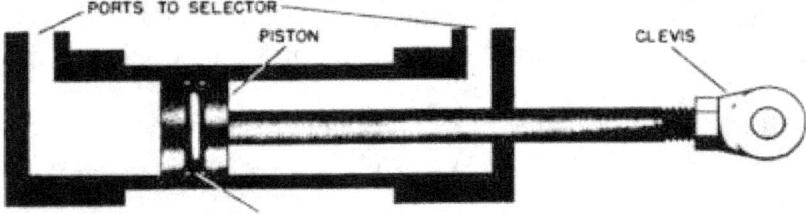

Figure 42.—Typical actuating cylinder showing end caps and clevi* bolt.

rod on one end only. Notice that the cylinder consists of a cylindrical housing containing one port opening into each end of the cylinder which alternates as inlet and outlet for fluid, depending upon the direction of flow from the selector valve.

End Cap and Clevis

In disassembling and assembling cylinders, it is necessary to understand the importance of the clevis and end caps.

The swivel-end cap assembly, located at the top of the end cap, may vary in design according to the type and installation of the cylinder. This unit should be periodically inspected for the purpose of lubrication and for examination of bushings and mounting.

The clevis bolt adjustment on most cylinders determines the location of the linkage travel, and must be adjusted accordingly. Incorrect location of travel will cause improper function of the unit which the cylinder is operating.

Piston-Head Seals

The seal on the piston head is shown in figure 42 as an O-ring, and it may be well at this point to consider a few pointers which will aid you in working with this type of seal.

Suppose that you were confronted with a piston, and that you had no manual at hand to explain its features. How would you determine the size of seal required in the piston? Here is a rule which should help you solve such problems:

Plus ten percent above, minus ten percent wide.

The seal extends approximately ten percent of its cross-sectional diameter above the edge of the groove, and has about ten percent overall clearance with the side of the groove. This clearance, as covered in our discussion of seals and packings, was % 4 -inch to V r2 -mch, but the ten-percent rule may be applied in emergencies.

Some types of cylinders will be found equipped with chevron seals. It will be recalled that this type of seal must be installed with the lips facing the flow and pressure because they seal in one direction only. It is now apparent that two or more chevron seals will be required in actuating cylinder piston-heads, with the lips of each set of seals facing the pressure on their respective side of the piston head. When installing these seals, make certain that female former rings are first placed on the piston head. Then set the lubricated seals in place one at a time. If more than one is used on a side, make sure of proper and secure seating. After inserting male formers, take care not to over-tighten the retainer nut. A good test for proper torque is to hold one seal lightly while turning the other seal. In other words, to prevent binding: "Hold one and turn one."

The same procedure is followed when installing seals around the end-cap shaft as is applied to seals on the piston head.

Certain pistons also have a felt wiper-ring on the inside of the end cap and around the shaft to guard against the entrance of dirt into the cylinder. Do not, however, put the entire task of dirt-elimination on the wiper-ring. Dirt is the greatest enemy of hydraulics, and a good mechanic will use that tried and true element, elbow grease, in daily inspections of exposed shafts.

Air in hydraulic cylinders—in fact, air in any hydraulic device or line—may result in improper, erratic operation'of the system. It is important, therefore, that all hydraulic devices be bled free of air after installation in the system.

Actuating Cylinder Adjustments

Adjustments on actuating cylinders usually are made at the clevis. Whenever it becomes necessary to remove the clevis, it is highly important that the number of turns required for such a removal be counted and MARKED. A very slight difference in the length of
mechanical linkage will invariably cause malfunction of the unit.

Another important item to bear in mind is the difference in effective working areas on the two sides of the piston head. The side to which the shaft is connected is obviously the shaft side of the cylinder. The opposite side is known as the blank side.

Suppose the blank side of the piston head has cross-sectional area of three square inches, while the cross-section of the shaft is one square inch. What would be the ratio of the amount of force developed by the blank side as compared with the shaft side of the cylinder?

It can easily be understood that an actuating cylinder may be so installed that the blank side carries the greater load, although no standard rule exists. The vital point to remember is that all actuating cylinders must be reinstalled in exactly the same positions as they occupied before removal.

As previously stated, single-acting cylinders are units containing a spring return. This unit has but one fluid port, located near one end of the cylinder. Figure 43 illustrates a single-acting, spring-loaded actuating cylinder which is used where a single stroke is required for unit-actuation or where the spring may provide emergency operation.

By releasing hydraulic pressure holding the spring
SINGLE-ACTING CYLINDERS

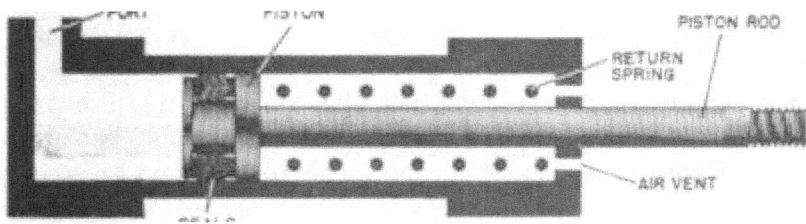

PORT
PISTON
SEALS

Figure 43.—Single-acting spring-loaded actuating cylinder.

compressed, the piston of a single-acting cylinder extends or compresses depending upon the side of the piston in which the spring has been installed.

We learned in chapter 2 that when we refer to pressure, we deal with the amount of force acting upon one square inch. Using that formula, and assuming that both sides of the piston are exposed to equal amounts of pressure, the ratio in our example would be three to two—three on the blank side and two on the shaft side. That is because the area of the shaft side subject to pressure is reduced to two square inches by the shaft itself.

The side of the piston opposite the fluid port is vented to the atmosphere, but is in contact with a coil spring. When fluid is introduced into the cylinder through the port, the piston moves toward the end of the cylinder opposite the port, and compresses the spring. When the fluid pressure is released, the spring expands and returns the piston to its original position. Fluid is forced through the port and back to the reservoir by the return travel of the piston.

PIN-PULUNG ACTUATING CYLINDERS

Pin-pulling actuating cylinders, termed double-acting balanced cylinders, usually are quite small and varied in design. A typical unit, which will be discussed in detail in a later chapter, is shown in figure 44. Notice that both sides of the piston shown have equal areas due to the fact that shafts of the same size are attached to

PISTON

SEAL

Figure 44.—Typical pin-pulling actuating cylinder.

either side of the piston. This type of actuating cylinder is extensively used in automatic pilots.

DASHPOT ACTUATING CYLINDER

When a plane leaves the deck of a carrier, it is most essential that its heavy landing gear be quickly raised. Naturally, it will require greater force to lift the gear than to lower it, but at the same time, the gear must be prevented from lifting right on up through the wing.

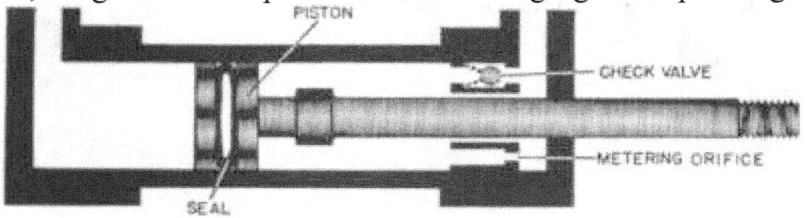

Figure 45. Dashpot actuating cylinder.

Actuating cylinders have been designed for just such purposes—to raise the gear and yet prevent damage to units. Figure 45 illustrates the dashpot actuating cylinder which contains a check valve designed to restrict flow in one direction while allowing free flow in the opposite direction.

Dashpot cylinders, however, are being replaced by re-strictors, restrictor checks, or orifice checks installed either in cylinder fittings or in lines or ports related to the action being controlled.

Suppose that in another situation a provision was required to prevent the gear from being too rapidly lowered. To take care of the junction, a restrictor would be installed in the UP pressure line or in the return line of the gear being actuated. It can be readily seen that return flow rather than pressure flow would thus be restricted, because a restrictor in the DOWN pressure line would be insufficient to prevent the gear from barging down.

SHAFT OIL PICKUP CYLINDER

The shaft oil pickup cylinder, shown in figure 46, has both ports on a special mounting bracket which directs flow through the shaft to either side of the piston. The pressure and return lines of flexible hose move back and forth as the piston responds to the action of the hydraulic fluid which they transmit.

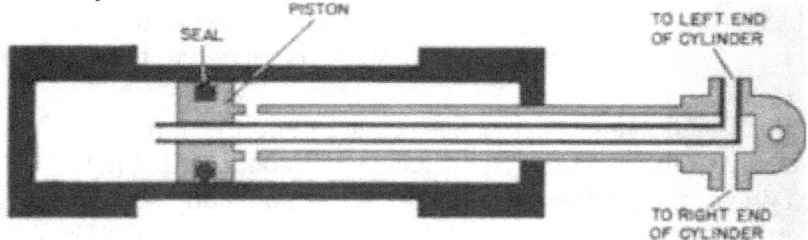

Figure 46.—Shaft oil pickup cylinder.

CYLINDER LINES AND FITTINGS

Most actuating cylinders are connected to working lines by flexible hose. It is highly important, following installation, to see that stripes on such hose run in straight lines without twist. Since hose tends to straighten under pressure, torsion or twist may cause the hose to shear off in the fittings, or the nuts holding the fittings to loosen.

Never stretch a hose too tightly between fittings. Five to eight percent of the total length of hose must be allowed as slack to provide freedom of operation under pressure. A hose will expand under pressure and may pull loose at its connections if slack is not provided.

Maintain a minimum bend radius of nine times the outside diameter of the hose.

Care should also be taken to make certain that flexible lines do not become frayed by rubbing against nearby parts of the airplane when the cylinder moves.

Swivels and swivel fittings are used to connect actuating cylinders to working lines where it is desirable to eliminate flex lines. Swivel fittings are used to attach lines to cylinder ports. Swivels provide flexibility for the lines.

Lines attached by swivels and swivel fittings should be inspected regularly for scoring, bends, and leaking.

QUIZ

1. What is the purpose of actuating cylinders?
2. Of what parts does a typical actuating cylinder consist?
3. For what purpose should the swivel-end cap assembly be periodically inspected?
4. By what rule in an emergency may the size of seal required in a piston be determined?
5. How must chevron seals be installed in actuating cylinders?
6. What is the greatest enemy of hydraulics?
7. What effect might air have in hydraulic cylinders?
8. What precautions should be observed when it is necessary to remove the clevis?
9. What distinguishes single-acting actuating cylinders from double acting cylinders?
10. What type of actuating cylinder is extensively used in automatic pilots?
11. What device in the dashpot actuating cylinder designed to lift landing gear prevents the wheels from being lifted right on up through the wings?
12. Why is it necessary to check stripes on flexible hose used in hydraulic systems to see that the hose is straight?
13. What percent of the total length of a hose should be allowed as slack to provide freedom of operation under pressure?
14. What bend radius is recommended for flexible hose installations?

15. For what should lines attached by swivels and swivel fittings be inspected?

ACCUMULATORS PURPOSE

You undoubtedly have read many anecdotes of "immortal" baseball pitchers who were able to call upon some hidden reserve power which enabled them to "bear down in the pinches." Just what constituted that intangible source of strength is open to conjecture, but in the hydraulic system such an emergency supply of power is also present. We will discuss this extra source of hydraulic power in this chapter.

When the baseball pitcher faced the weak end of his opponents' batting order, he breezed along, storing up and holding in reserve the strength and cunning needed for use against the heavy hitters. Then, when his ordinary pitches would have been plastered against the ballpark fences, he dipped into his hoarded reserve and came

up with the necessary "stuff" on the ball to keep his team in the game.

In the hydraulic system, we might term the pump the "pitcher," and the pressure accumulator the reserve power standing by in case of emergency. We have discussed actuating and energizing units, and have seen that the reservoir in an aircraft hydraulic system is a storage unit where the hydraulic fluid for the system is stored to supply fluid for the energizing units. In this chapter we will consider another type of storage unit— the accumulator.

The purpose of the hydraulic accumulator is to store hydraulic fluid under pressure. In accomplishing this, the accumulator performs the following functions:

1. Maintains pressure in the pressure manifold by storing energy in the form of fluid under pressure. This function enables the accumulator to supplement the power pump when it is under a peak load.

2. Supplies a limited amount of fluid under pressure to actuating units when the power pump fails to operate.

3. Dampens the pressure surges which may be caused by the pulsating fluid delivery from the power pump.

4. Absorbs fluid shocks, such as those occurring when the pressure regulator seals the pressure manifold and directs the fluid to the reservoir.

5. Prevents too frequent cut-in and cut-out of a pressure regulator. This function reduces

wear on the pump and the pressure regulator.

Accumulators are not used in all aircraft hydraulic systems. The accumulator is used in conjunction with a pressure regulator (unloading valve) in one type of direct pressure system as a means of maintaining a constant high pressure at control valves, while pressure is relieved from the power pump when actuating equipment is not in use.

Accumulators are also in other types of systems. In fact, when an aircraft hydraulic system is designed, an accumulator can be incorporated into the system at the discretion of the designer.

Loaded accumulators are very dangerous devices and must be treated with respect. The air load must always

BE RELEASED BEFORE DISASSEMBLY. This should be done
by depressing the air valve pin—NOT BY UNSCREWING
THE VALVE.

CLASSIFICATION

There are three main types of pressure accumulators —the diaphragm type, the bladder type, and the piston type. Generally, the diaphragm, or spherical, type accumulators are manufactured in 5-inch, 7%-inch, and 10-inch inside diameter sizes, depending upon the volumetric requirements of the system.

Diaphragm-Type Accumulator

One form of diaphragm-type accumulator, illustrated in figure 47, is made up of two forged steel hemispherical shells which are screwed or bolted together to form a complete sphere. The top half has a fitting for connecting the unit to the system. Fluid is pumped through this connection, thus forcing the diaphragm down and compressing the air in the lower half.

A screen, installed across the inside of the port, prevents the diaphragm from rupturing when the accumulator is preloaded with air and when the fluid is drained into the system.

The lower half of the accumulator is equipped with an air valve for charging the unit with compressed air at a predetermined pressure. During the air-charging operation, the diaphragm unfolds and contacts the inner surface of the upper half of the accumulator.

Mounted between the two halves is a synthetic rubber diaphragm that divides the tank into two chambers, one for air and one for fluid, as shown in the illustration.

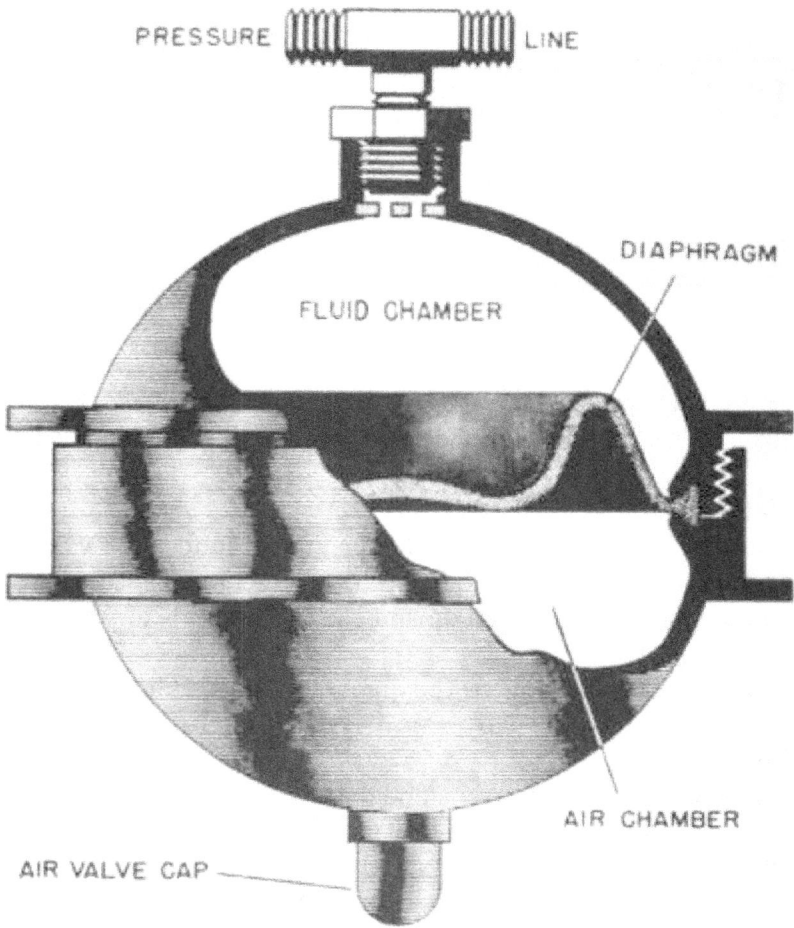

Figure 47.—Diaphragm accumulator.

Normally, the accumulator is mounted with the air valve down, thus placing the air chamber in the lower position and the fluid chamber in the upper position.

An initial charge of compressed air is placed in the accumulator. The pressure of the air forces the diaphragm upward. When the pressure of the fluid in the system becomes higher than the air pressure in the lower half, the fluid is forced into the top chamber, and there it forces the diaphragm downward, further compressing the air in the lower chamber.

When there is a peak load, the highly compressed air tends to force fluid back into the system. Also, if the power pump fails to function, the compressed air in the lower chamber pushes on the diaphragm and thus supplies a limited amount of fluid under pressure to operate some mechanism.

Bladder-Type Accumulator

Figure 48 illustrates another type of accumulator— the bladder type. Operation of this type accumulator is practically identical with that of the diaphragm type. The main difference is found in the shapes of the shell and the bladder.

In the bladder-type accumulator, the housing, or shell, is a hollow steel cylinder with rounded ends. This unit contains two diametrically opposed ports. A synthetic rubber bladder is installed through the lower port and sealed to the port by means of a cap. The top port is connected to the pressure system.

When the accumulator contains no fluid and the air charge is released, the bladder is approximately the same size as the accumulator shell. Therefore, the bladder is not stretched by air pressure.

The metal disk, located in the center of the bladder,

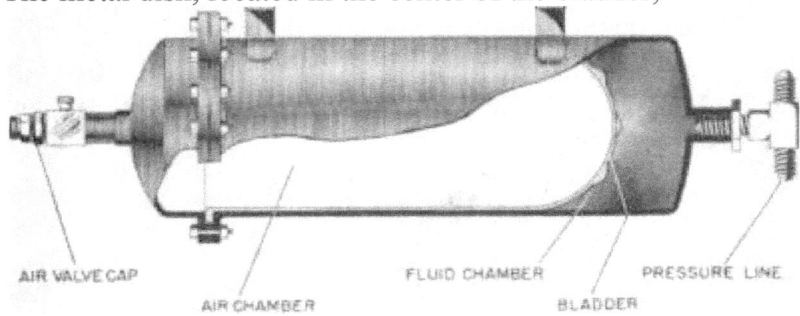

Figure 48. — Bladder-type accumulator.

prevents the bladder from extruding through the upper port when the fluid is drained from the system.

Piston-Type Accumulator

The piston-type accumulator is a cylinder in which a piston separates the two chambers. O-ring seals, as shown in figure 49, are installed on the piston to prevent leakage between the two chambers. Both ends of the cylinder are screwed on and use O-ring seals to prevent external leakage.

The piston in this unit acts in the same way as the diaphragm or bladder functioned in the previously discussed types of accumulators.

VICKERS ACCUMULATOR

The Vickers and Bendix accumulators are AN standard models, and are the type most commonly used in Naval aircraft hydraulic systems.

The Vickers accumulator consists of a seamless steel spherical container separated into an air chamber and a fluid chamber by a synthetic rubber diaphragm. The air chamber is charged with compressed air or nitrogen and is equipped with an air valve for recharging. The fluid chamber is connected to the pressure line of the system and is filled with hydraulic fluid at the system operating pressure.

Hydraulic fluid enters the sphere and compresses the air in the diaphragm until the pressure equalizes. The compressed air then acts as a cushion to dampen out any pressure surges in the system, thus relieving the un-loader valve from constant operation.

BENDIX ACCUMULATOR

The Bendix accumulator is composed of a shell assembly, bladder assembly, and cap assembly. It differs from the Vickers accumulator in that the shell assembly is not divided into two hemispheres, but is a spherical shaped housing inside of which the bladder assembly is installed.

The bladder is a synthetic rubber envelope which separates the air and oil chambers of the shell assembly. The cap assembly seals the open end of the bladder assembly and is held firmly against the flange of the bladder assembly by a retainer. An air valve in the'cap assembly furnishes a means of inflating and deflating the accumulator.

Air pressure fully extends the bladder against the walls of the shell. As hydraulic fluid is forced through the oil port, the stop end of the bladder is forced toward the bottom, further compressing the air in the bladder. When fully charged with oil, the bladder is U-shaped in the bottom of the accumulator.

The operation of the Bendix and Vickers accumulators is similar.

inspection and Maintenance

Accumulators should be visually examined for indications of external hydraulic-fluid leaks and then -should be examined for external air leaks by brushing the. accumulators with soapy water. The soapy water will form bubbles where the air leaks occur.

A high-pressure gage is nedessary to check the charge of air. The pressure-gage reading must be watched carefully. The last reading before the needle drops suddenly to zero is accepted as the accumulator air pressure.

In testing accumulators, first reduce the air preload to zero, then apply hydraulic pressure equal to one and one-half times the system pressure at the system port with the air valves loosened. After checking for leaks, apply the specified air preload to the air chamber side with the oil port open. Submerge the accumulator in hydraulic fluid and check for leaks.

Both high and low-pressure air valve cores are used on aircraft, but only the high-pressure type may be used on accumulators. Low-pressure cores have a rubber seat

FLUID CHAMBER'
AIR CHAMBER
AIR VALVE CAP
Figure 49.—Piston-type accumulator.

while the high-pressure cores have a lead seat and the raised letter H on the head of the stem. A new air-valve core and crush washer should be installed each time the accumulator is disassembled, and whenever air leakage is evident.

QUIZ

1. What is the purpose of the hydraulic accumulator?
.2. Name the five functions performed by the hydraulic accumulator.
3. What essential precaution must be observed before disassembling accumulators?
4. Name the three main types of pressure accumulators.
5. How are accumulators tested for external air leaks?
6. What causes frequent cut-in and cut-out of automatic pressure regulators?
7. How might loss of air pressure be remedied?
8. What should be done to remedy line hammer?
9. What reaction or trouble would be caused by a stoppage in the oil screen?
10. How should a damaged diaphragm be remedied?

CHAPTER 10
LOCKS AND SEQUENCE VALVES PURPOSE

On most military aircraft, the landing gear is fully retractable, and all modern carrier-based planes have folding wings. It is obvious that some provision must be made for locking those units in a fixed position—especially the extended or "down" position of the landing gear, and the spread position of the wings.

The discussion before us will consider those provisions which have been made for locking units such as landing gear and wings. These will be found embodied in various designs of hydraulically actuated locks and
locking pins especially designed for the purpose.

Some types of landing gear down-locks are operated by the landing-gear actuating cylinder. These locks include detent-ball types, spring-loaded locking pin, and the horseshoe, or J-hook.

WING-LOCK CYLINDER
Now let's consider the wing-lock cylinder which performs the actual work of securing the

wings in the spread position described above.

When the selector valve is placed in the locked position, the flow of fluid is directed to the sequence valves. When the wings are fully extended, the sequence valve plungers are depressed, allowing the fluid to flow into the wing-lock cylinders and to extend the locking pins.

After the hydraulic pins are extended, the pilot secures the mechanical lock which holds the hydraulic locking pins in the extended position. (The mechanical lock is operated by a spring-loaded toggle linkage accessible to the pilot).

To unlock the wings, the pilot first unlocks the mechanical locks and then places the selector valve in the unlock position. This causes the reverse of the above action to take place, thereby unlocking the wings.

Wing-folding systems may have one of several types of locking pin actuating cylinders. The common type is a unit with a piston shaft through each cylinder cap for the simultaneous operation of two locking pins. Another style is the snap action type which uses a spring and detent balls to furnish a quick and positive locking and unlocking action. This cylinder has a specially-constructed piston which acts as a sequence valve.

A third type is one with a sequence valve incorporated in the end cap. The sequence valve plunger pin is contacted by the lock-actuating cylinder piston in the unlocked position, thus opening the sequence valve.

SEQUENCE VALVES

The function of sequence valves is to cause one hydraulic action to follow another in a definite order or sequence. This unit is also called a "timing valve" or "load and fire" check valve.

Sequence valves differ in details of construction and design, but are similar in operating principles. The unbalanced type is the one most commonly used.

The sequence valve is essentially a bypass valve which is operated automatically. It consists of an outside housing in which there are two ports, a poppet (or ball) and seat, two return springs, and the necessary seals to prevent leakage. In one end of the housing is an external device for operating the ball or poppet. When this device is actuated, it in turn thrusts the ball or poppet off its seat.

The sequence valve is mounted in such a position that motion of some part of the mechanism—the landing gear, for example—will cause the unseating device to be moved when the mechanism reaches the end of its stroke. A good rule to remember when working with units employing sequence valves is that the working line under

PRESSURE FROM THE SELECTOR VALVES GOES TO THE FIRST UNIT TO BE OPERATED, THEN THROUGH A SEQUENCE VALVE TO THE SECOND UNIT TO BE OPERATED.

Sequence Valve Adjustment

On the wing-fold systems of the Model F6F aircraft, an adjustable contact pin is attached to the outer wing panel, and contacts the sequence valve plunger when the wing spreads. The contact pin should be adjusted until the free flow of fluid is allowed when the wing is fully spread. If the sequence valve plunger is adjustable, it should be similarly adjusted.

QUIZ

1. What must be done to unlock the wings of a plane having a wing-folding system?

2. What is the common type of locking pin actuating cylinder - found in wing-folding systems?

3. What type of locking pin actuating cylinders in wing-folding systems uses a spring and

detent balls to furnish quick and positive locking and unlocking action?

4. What is the function of sequence valves?

5*» What type of sequence valve is most commonly used?

6. How should the contact pin attached to the outer wing panel in the wing-fold system of an F6F be adjusted?

CHAPTER 11

RELIEF VALVES PURPOSE

When human blood pressure soars, and a man is about ready to "blow his top," the traditional remedy prescribes a cooling-off period commonly known as "counting to 10." Sometimes such a measure proves successful.

But whoever heard of an aircraft pilot setting his plane down for a ten-second pause when the ship threatened to get out of hand? It just isn't done! Yet something must be provided to keep mechanical tempers in line. For any type of imprisoned power must be controlled—whether that power be Aladdin's genii held in check by a cork in a bottle, or hydraulic pressure governed by relief valves.

Pressure relief valves are placed in the hydraulic system to keep hydraulic pressure at predetermined safe values and to control it when it threatens to get out of

hand. The relief valve is really a safety valve that protects the system from the damage which might follow as a result of excessive pressure.

GENERAL FUNCTION OF RELIEF VALVES

Relief valves consist of a two- or four-port housing, a ball or cone held on its seat by a strong spring, and an adjusting screw. A typical relief valve is shown in figure 50.

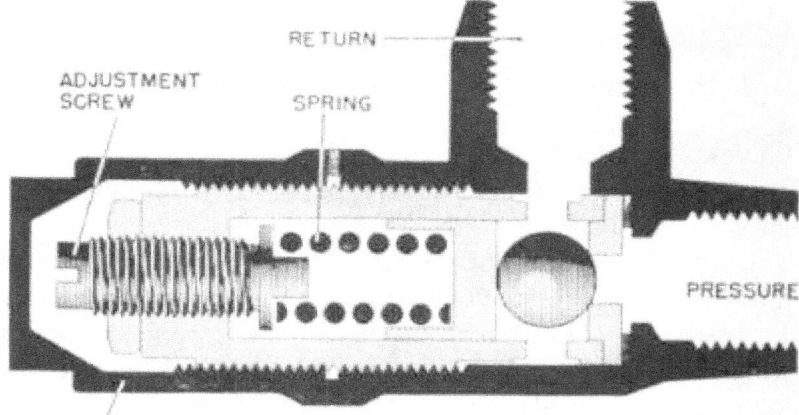

COVER

Figure 50.—Typical relief valve.

As may be seen in figure 50, fluid cannot pass from the pressure port to the return port as long as the ball is held on its seat. When pressure at the pressure port becomes sufficiently great to overcome the force of the spring, it moves the ball off the seat, allowing fluid to escape to the return line through the return port.

CLASSIFICATION

Two types of hydraulic pressure relief valves are in general use in aircraft hydraulic systems. In each case, the basic design involves a spring-loaded valve arranged so that it automatically opens to relieve the system

pressure when fluid pressure acting on one face of the valve becomes sufficient to overcome spring pressure applied to the opposite face.

These two types of relief valves include main system relief valves which safeguard the system against excessive pressure in the event of regulator failure; and thermal relief valves, designed to compensate for thermal expansion in various unit systems.

MAIN SYSTEM RELIEF VALVE

Main system relief valves usually are adjusted to relieve above system pressure, system pressure being the cut-out of the regulator or maximum setting of the pressure-control system.

The main relief valve in the F8F-model aircraft, for example, is connected to the main pressure line immediately following the main system filter. It is of spring-loaded poppet-type, unbalanced construction, and is adjusted to open if pressure in the main line builds up to 1,750 p. s. i. During normal operation, the valve will remain closed, but, if through failure of the unloader valve to operate properly, the pressure rises to 1,750 p. s. i., the spring-loaded poppet is unseated, allowing fluid to bypass through the valve to the reservoir.

Main system relief valves may be subdivided into balanced and unbalanced types.

Balanced Relief Valves

Balanced relief valves are used as system relief valves in some aircraft, but are more commonly employed to control pressure in autopilot systems.

By balanced we mean that fluid under pressure is acting on both sides of a piston which has equal areas on each side. A small metering hole is drilled through the piston to effect the metering of fluid from the high-pressure side to the low-pressure side of the balanced piston when the relief ball is unseated. Therefore, the high-pressure on one side of the balanced piston will force the

piston toward the low-pressure side, causing the piston poppet to unseat and allow fluid to flow to the return.

When excessive pressure to the return port has been relieved, the spring on the opposite side of the balanced piston will force the piston to move in the opposite direction, thus seating the piston poppet and closing the return.

Unbalanced Relief Valves

Unbalanced relief valves are of either ball or poppet type, and in describing operations reference to unbalanced relief valves will embrace either category.

The poppet in unbalanced relief valves is held on its seat by a spring which is maintained at a desired tension by an adjusting screw. Pressure building up in the system is also building up beneath the poppet. When the pressure becomes sufficiently great to overcome spring tension, the poppet will unseat, allowing excessive fluid to flow to the return. When the force overcoming spring tension is gone, tension of the spring will then close the return.

THERMAL RELIEF VALVE

The operation and adjustment of thermal relief valves is similar to that of the main system unbalanced type. On the landing gear of the F8F-model plane, for ex-

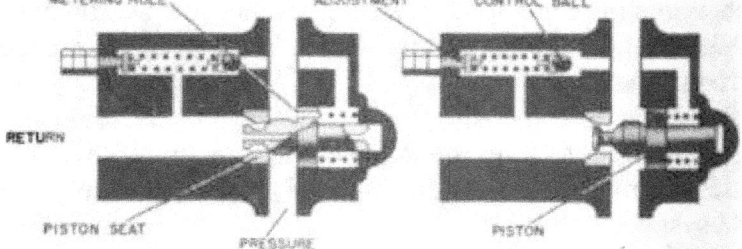

NORMAL OPERATION RELIEVING PRESSURE
Figure 51.—Balanced relief valves in open and closed positions.

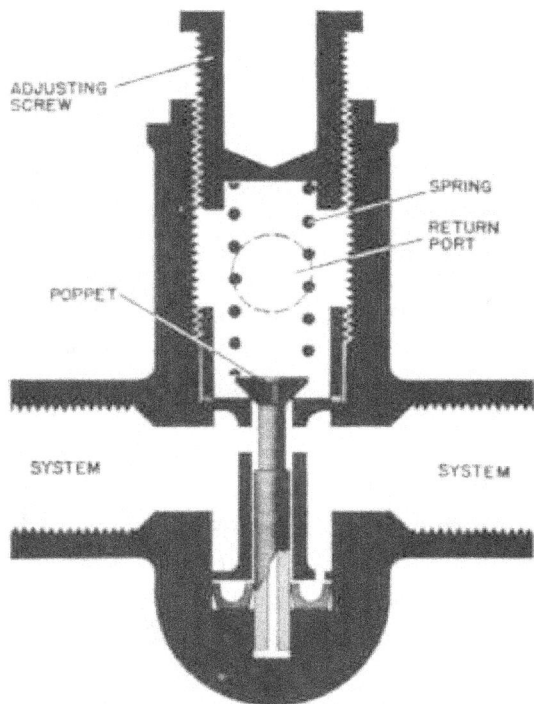

Figure 52.-Unbalon«d relief valve.

ample, a relief valve set to open at 2,000 p. s. i. is connected to the door-line of the landing gear system to relieve the system of excessive pressure due to heat expansion or to other causes. This valve is mounted on the forward port side of the firewall below the fuselage centerline.

This valve has been installed on all F8F aircraft to prevent damage to cylinders formerly resulting because of cylinder-swelling when engine exhaust gases heated the liquid. To avoid this condition, a thermal relief valve was installed with a setting of 2,000 p. s. i.

RELIEF VALVE ADJUSTMENT

In most cases, relief valves can be adjusted without removal from the plane, although it is preferable to adjust them on a test stand.

When adjusting relief valves in a system, adjust but one valve at a time. Back out the adjusting screw of the valve to be adjusted, and turn the adjusting screws of all other relief valves inward. Begin with the valve having the highest opening pressure, and progress in descending order to the valve with the lowest opening pressure.

Turn the adjusting screw inward until the pressure indicated on the gage is that required. Take several readings and consider the average as the setting of the valve. If there is a power-control valve in the system, it must be held closed and adjusted last of all.

A simple rule to remember in making pressure adjustments on hydraulic units is THE HIGHER THE COMPRESSION OF A SPRING BECOMES, THE HIGHER THE PRESSURE REQUIRED TO OVERCOME IT.

QUIZ

1. What device is provided in the hydraulic system to keep hydraulic pressure at predetermined safe valves and to control it when it threatens to get out of hand?

2. Of what does a typical relief valve consist?

3. What device within a relief valve prevents fluid from passing from the pressure port to the return port?

4. Name the two main types of relief valves.

5. What are the two types of main system relief valves?

6. What is the purpose of the main system relief valve?

7. For what pressure is the main system relief valve set in the F8F?

8. By what means is the metering of fluid from the high-pressure side to the low-pressure side of the balanced piston when the relief ball is unseated in the balanced relief valve effected?

9. In unbalanced relief valves, how is the desired tension of the spring which holds the poppet on its seat attained?

10. At what pressure is the thermal relief valve on the landing gear of an F8F set to open?

11. For what purpose is the thermal relief valve installed on the landing gear of the F8F?

12. What simple rule should be remembered when making pressure adjustments on hydraulic units?

CHAPTER 12

AUXILIARY UNITS DUMP VALVES

The kitchen of a home functions principally with such major units as a stove, refrigerator, and sink. But whoever prepares meals in a kitchen would be severely handicapped unless a large variety of auxiliary units— knives, spoons, choppers, graters, and mixers—were included in various drawers and cupboards. It takes a heap of odds and ends of gear to make up an efficient kitchen.

In the same way, no aircraft hydraulic system is complete when only such units as the pump, reservoir, and actuating cylinders are hooked up and functioning. A multitude of assorted auxiliary units must not be overlooked or ignored if the hydraulic system is to perform its vital functions smoothly and flawlessly. Therefore, we need to consider those auxiliary or "secondary" units —how they work and why they are used.

VALVE A

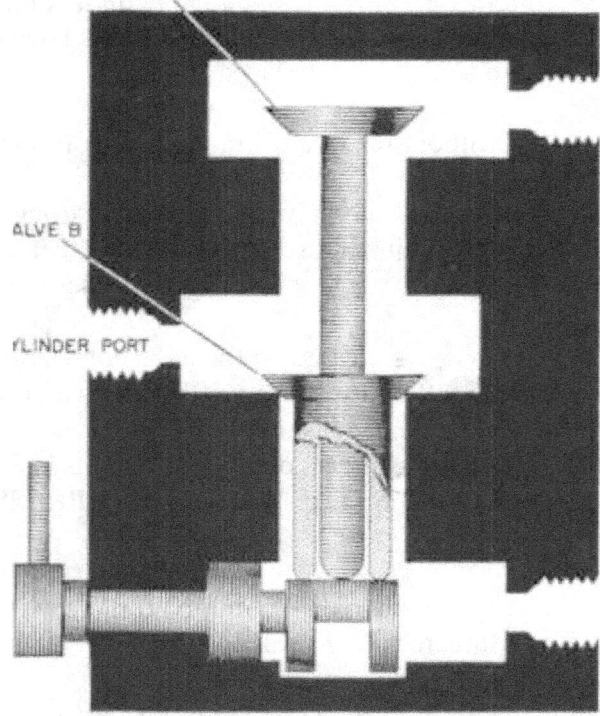

VALVE

CYLINDER PORT
FROM SELECTOR
PORT
Figor# 53.—Typical

A type of emergency valve in use on certain naval aircraft is the dump valve which, as the name implies, simply "dumps" or releases hydraulic fluid. Usually, dump valves are used in connection with emergency operations of landing gear, bomb-bay doors, and as system pressure dumps.

Dump valves are installed in the landing-gear unit, for example, so that the wheels can be lowered, in event the hydraulic actuating system should fail completely. The majority of dump valves are very similar in construction, so that a description of one will apply to all. A typical emergency dump valve is diagrammed in figure 53.

In this valve, an aluminum-alloy housing provides con-
nections for lines leading from the selector valve to the landing-gear actuating cylinder and thence back to the reservoir. Within this housing is a coaxial valve which is really a "valve within a valve." A cam is located directly in line with the valve, and is equipped with an operating lever.

The valve is operated by turning the lever which acts against the cam, causing the high point of the cam to force valve A upward. This permits the fluid to flow from the selector valve to the actuating cylinder. The unloading action occurs when the selector handle is moved in the opposite direction. The fluid is then free to flow overboard, through valve B. This is called the "dumping action." When this occurs, restraining action no longer is applied to the gear.

CHECK VALVES

The purpose of check valves is to allow free fluid flow in one direction while prohibiting flow in the opposite direction. Check valves may be used to trap pressure in part of the hydraulic system, and may be installed in the pressure or return manifold. Several variations of check valve types include orifice check valves and

BYPASS CHECK VALVES.

The checking device of this unit may be a ball, a cone, or a flapper plate, held on its seat by a light spring.

PASSAGES AROUND FACE OF CONE EQUAL AREA OF PORT A

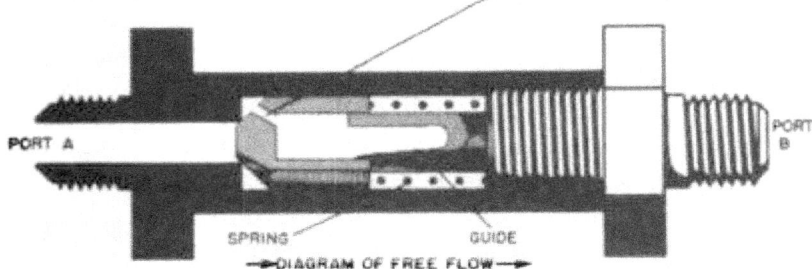

Figure 54.—Typical check valve.

This device is ground to fit the seat perfectly to prevent leakage.

In the typical check valve illustrated in figure 54, fluid entering port A will unseat the cone and emerge from port B. When the flow of fluid ceases, the spring returns the cone to its seat, thus trapping the fluid which has passed through the unit. The direction of free flow through the valve is indicated by an arrow stamped on the housing.

Orifice Check Valves

Orifice check valves are used in hydraulic lines to allow free fluid flow in one direction and restrained flow in the other. This valve is used to retard the operation of flaps, landing gear, and similar units, in one direction.

As may be seen in figure 55, this unit is essentially a

RESTRICTED FLOW   FREEFLOW

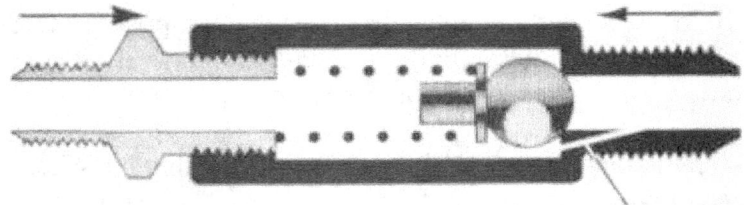

DRILLED PASSAGE

Figure 55. Orifice check valve in closed position.

check valve with a small hole drilled in and through the seat. Even when the valve is in the closed position, the orifice allows a small amount of fluid to flow back through the unit. With this fluid flow in the normal direction, the valve acts in* the r same manner as an ordinary ball-check valve.

Orifice check valves are used in some wing-flap systems to prevent air pressure from pushing the flaps upward too rapidly. When used in landing-gear systems, these valves delay extension of the gear by counteract-

ing the natural tendency of the weight of the gear to force it down too rapidly.

MANIFOLD OR JUNCTION BLOCKS

The manifold block acts as a junction box for the hydraulic lines, and is designed to save weight and space by eliminating many fittings. This also tends to lessen the danger of leakage.

Manifold blocks are located at branch-off points of pressure lines to selector valves; working lines to cylinders, and return lines from selector valves.

Construction of manifold blocks is simplicity itself. The units are machined, drilled, and threaded for the designed purpose.

The Kenyon Manifold Block incorporates check valves and thermal relief valves, thus further economizing in weight and space.

RESTRICTORS

To restrict means to slow, or hamper, and the purpose of restrictors in hydraulic systems is to limit the rate of flow in both directions in a line. In so doing, these units cause the mechanism being actuated to move more slowly.

An orifice and a variable restrictor differ only in construction. The size of an orifice is usually fixed, whereas the size of the opening in a variable restrictor may be changed. An orifice is primarily a fitting containing a small passage. Fluid entering one end of the fitting must traverse the small passage before flowing from the opposite end. The housing of a variable restrictor, illustrated in figure 56, has two ports and an adjustable needle valve. The size of the passage through which the fluid must pass may be adjusted by screwing the needle valve in or out.

In some cases, if heavy units—such as landing gear— were allowed to fall free to extended positions, structural damage would result. Elimination of this hazard is

ADJUSTABLE NEEDLE VALVE

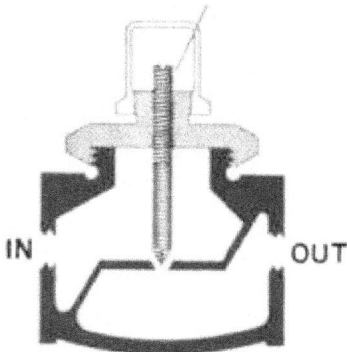

Figur. 56-VariobU rMtrictor.

effected by installation of a restrictor in the UP line to restrict the flow of fluid emerging from the actuating cylinder, thus causing the gear to move more slowly.

FLOW EQUALIZERS

The flow equalizer, also called the "flow divider," is used in unit systems where simultaneous operation of two or more actuating cylinders is desired.

There are two types of flow equalizers—the PISTON type and the gear type. Since the gear-type unit is used principally in naval aircraft, that model will be considered in greatest detail in our discussion.

Gear-Type Flow Equalizer

The gear-type flow equalizer consists of an aluminum-alloy housing or body containing two gear-type hydraulic motors of equal displacement. A common drive shaft, maintaining synchronization at all times, connects the two sets of gears. Two perfectly balanced regulating valves, one for each set of gears, are also contained within the housing. These valves provide an equal flow of fluid from the outlet ports of the flow equalizer to the actuating cylinders. The flow equalizer is provided with

three ports—one outlet port connected to the actuating cylinder control valve, and two outlet ports, one for each side of the flow equalizer connected into the left and right actuating cylinders.

Fluid from the selector valve enters the inlet port of the flow equalizer, passing to both motors. As these motors revolve at equal speeds, the flow is evenly divided to the two working lines. The thermal relief valves in

SPLINES KEY THE TWO THERMAL

SET OF GEARS Figur« 57.-Gear-type flow equalizer.

the housing compensate for thermal expansion in the working lines.

The flow equalizer operates accurately under discharge port pressure differentials up to 1,400 p. s. i. Equalization in reverse flow is not nearly so accurate.

CONTROL BOOSTERS

As aircraft speeds and sizes increase, it becomes more and more difficult to manually operate aircraft controls. The strength and endurance of the ordinary man is not great enough to manipulate controls at high speeds for extended periods of time. Therefore, to aid the pilot in moving plane controls, hydraulic control boosters have been devised.

Booster operation is simultaneous with mechanical manipulation. When the surface control is operated by the pilot, contact points move a valve, and fluid is directed to the actuating cylinder. Thus the control surface is moved by the hydraulic boost pressure.

One end of the actuating cylinder is attached to the stationary structure, while the other end (the piston shaft) connects to the moveable control. Whenever the pilot moves the controls, the proper poppets are opened in the selector valve.

In the event that fluid should become locked in the actuating cylinder and thus lock the control surface, an emergency bypass valve, which bypasses fluid from the actuating cylinder to the reservoir, can be operated by the pilot from the cockpit.

GAGES AND GAGE-SNUBBERS

Generally, all pressure gages used in hydraulic systems, give a gage reading. Calibrated to indicate p. s. i. pressure above atmospheric pressure (14.7 p. s. i.), the luminous gage-dial is located in the cockpit, or on the flight engineer's panel in some multi-engine planes. The purpose of the gage is to measure hydraulic pressure in pounds per square inch.

The hydraulic gage consists of a curved tube closed at one end and open to system pressure at the other. The tube tends to straighten out when internal pressure is built up. A mechanical linkage is connected and calibrated between the closed end of the tube and the indicating pointer. Thus the hydraulic pressure is translated directly to an amount of rotation of an indicator.

The irregularity of impulses applied to the hydraulic system by some engine-driven power pumps causes the pressure-gage pointer to oscillate violently. This is disconcerting to the observer, and renders accurate reading not only difficult but often impossible. Pressure oscilla-tions and other sudden pressure changes existing in hydraulic systems will also affect the delicate internal mechanism of gages and cause either complete destruction of the gage or, often worse, partial damage resulting in false readings. A pressure-gage snubber is therefore installed in the line to the pressure gage. The purpose of this unit is to dampen the oscillations and thus provide a steady reading and a protection for the hydraulic gage.

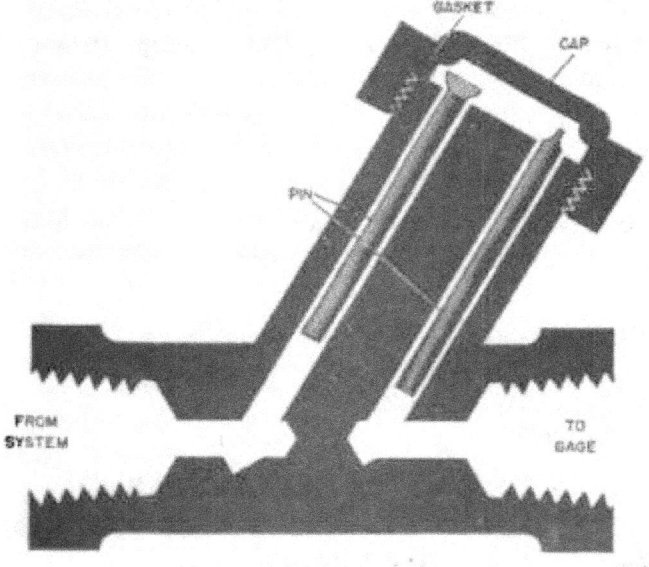

Figure 58.—Pressure-gage snubber.

• * j

The pressure-gage snubber shown in figure 58 consists of a two-port housing containing an orifice into which a floating plunger is inserted. This plunger tends to reduce the area of this orifice, thus increasing the dampening effect. The floating action of this plunger causes it to move back and forth in the orifice, thus making the small opening "self cleaning." This reduction in surges results in a steadier pointer indication and' a more accurate reading.

PRESSURE SWITCH

The function of a hydraulic pressure switch is to turn an electric switch on and off automatically in response to changes of pressure. In most installations using an electric motor for operation of a hydraulic pump, a pressure switch is installed instead of a pressure regulator to take the load off the pump when no units are being operated.

Cut-in and cut-out pressures in these units must be higher than the system regulator cut-in and cut-out pressures if the pressure switch is used with another system. This is because the auxiliary pump, in loading the accumulator, must not stop before the pressure regulator cut out. This keeps the engine pump unloaded and prevents overwork of any units in the power system.

The master switch for the electric motor is located in the cockpit. When the engines are running, the electric motor will be shut off, except during landings and take-offs.

ARRESTING HOOK

An hydraulically operated arresting hook is used on most naval aircraft assigned to carrier operation. On planes with conventional-type landing gear, the arresting hook is located on the lower tail end of the fuselage. On aircraft with tricycle landing gear, this unit is situated farther toward the nose, and usually is longer.

On carrier planes, such as the F4U-1, the hook is extended by hydraulic pressure as well as by its own weight. To retract the hook, hydraulic fluid enters the shaft side of the actuating cylinder. The hook is operated by the four-bank selector valve.

However, the valve operating the hook is somewhat different from the others as a pressure of 250 p. s. i. is required to hold the hook up. A back pressure of 400 p. s. i. is maintained in the UP working line by a slightly unbalanced' pressure poppet in this selector valve. A dashpot assembly prevents the hook from bouncing off the deck when making a landing. To further streamline the aircraft, the arresting hook is retracted into the fuselage.

HYDRAULIC FUSES

Hydraulic fuses are new units added to later models of naval aircraft for the purpose of preventing the loss of hydraulic fluid from broken lines. Fuses are strategically located throughout the hydraulic system.

So long as the system operates normally, the fuse remains open. When a line protected by a fuse is broken, however, the fuse will act as a safety shut-off after a measured quantity of fluid has passed through this unit.

Let us suppose that 40 cubic inches of fluid are required to actuate a cylinder and that the working line between the fuse and the actuating cylinder has become broken. After 40 cubic inches of fluid have passed through the fuse, the unit will close and shut off the flow.

The two types of fuses are the quantity measuring type and the return flow type.

ELECTRICALLY OPERATED HYDRAULIC UNITS

Electricity is sometimes used to operate selector valves and shut-off valves. These valves may be automatically controlled by pressure switches, or thermostats, or they may be controlled by a toggle-switch in the cockpit.

Poppets in these valves are operated by solenoids, or small electro-magnets. Solenoids are employed where only small movements are required, and are commonly used as relays, especially where high currents must be handled remotely. Basically, solenoids consist of a winding around a moveable iron core. When the coil is energized, the core becomes magnetized and moves into the coil in an attempt to center itself. When the coil is de-energized, spring action returns the core to its original position.

The solenoid-operated shut-off or stop valve is used in parts of the system remote from the cockpit. This eliminates the need for excessive mechanical linkage and saves weight. The same holds true for solenoid selector valves.

Solenoid Shut-Off Valve

To simplify the operational description of the solenoid-operated shut-off valve, it is presented in the schematic flow diagram in figure 59 as installed in a typical aircraft hydraulic system to provide a remote ON-OFF control of hydraulic fluid.

When the electrical circuit to the unit is open, the solenoid is off, the solenoid plunger is extended, and the valve is in its normal position with the shut-off poppet open, as shown in view A of figure 59. Fluid flow is permitted in either direction—from inlet to outlet or from outlet to inlet.

View A of figure 59 shows the electrical circuit open. The solenoid is off and the plunger is extended. The valve is in the normal position, permitting fluid flow in either direction. In view B, the electrical circuit is closed, the solenoid is off, and the plunger is retracted. The valve is in the operated position, and fluid is not permitted to flow from inlet to outlet. View C, identical with view B, shows the fluid flow permitted to pass from outlet to inlet.

When the electrical control switch is operated by the pilot, electric current is directed to the coils of the solenoid, converting that unit into an electro-magnet. The solenoid plunger is drawn into the core by magnetic force, pulling the lever up and allowing the shut-off poppet spring to force the poppet into a seated position, thereby affording a complete shut-off of fluid flow from inlet to outlet, as shown in view B. However, fluid is permitted to flow from outlet to inlet via the check valve, as shown in view C. So long as the switch remains ON, current will hold the solenoid plunger in a retracted position. When the switch is turned off, the solenoid plunger will be forced back out of the core by spring action against the lever, which again unseats the shut-

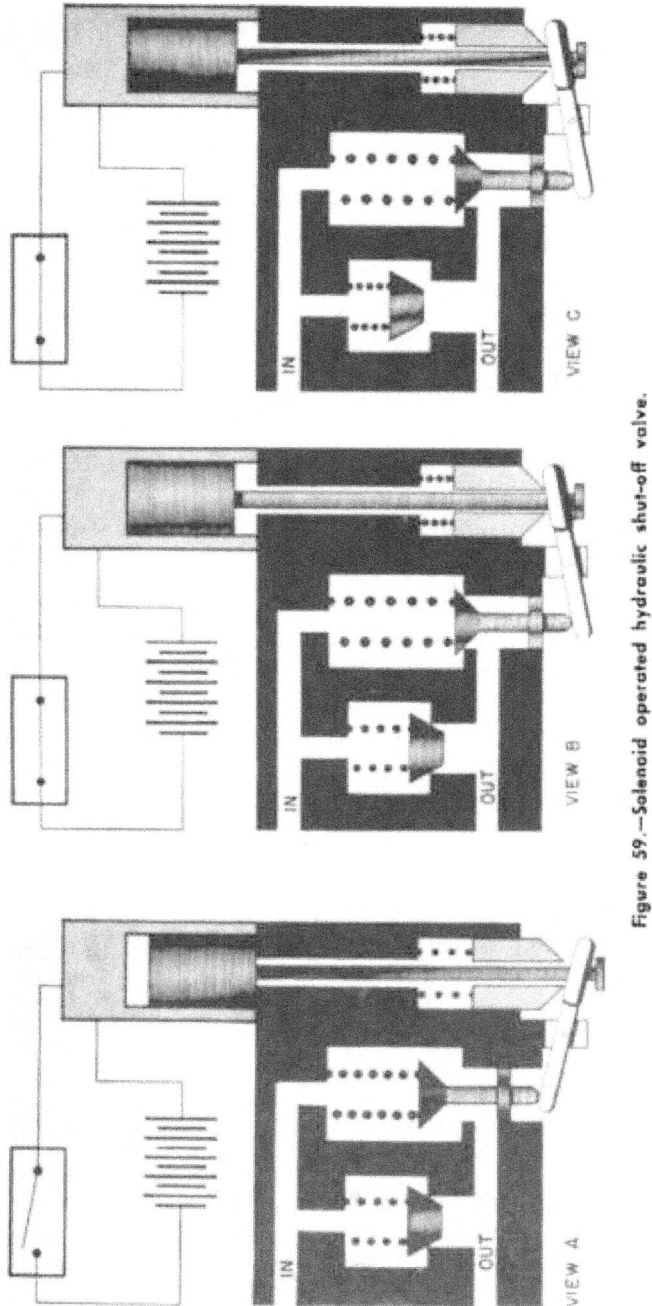

Figure 59.—Solenoid operated hydraulic shut-off valve.

off poppet assembly for a normal flow- *. of fluid.

    The solenoid has two electrical coils—a pulling coil and a holding coil—connected in multiple (or parallel) so that both receive current at the same time. As the electrical circuit to the solenoid is closed, its two coils retract the solenoid plunger by electromagnetic force, thereby pivoting the lever on the pin and allowing the spring to force the shut-off poppet into a seated position and shutting off the flow of fluid.

    As the solenoid plunger completes its inward stroke, it opens a switch which cuts out the main pulling coil, but leaves the holding coil in the circuit with sufficient power to hold the plunger in its full-in position. When the electrical circuit is again opened, the holding coil is also cut out and the solenoid plunger thereby released. The spring, acting against one end of the lever, pivots the lever back to its normal position and the shut-off poppet is forced open to allow normal fluid flow as shown in view A.

QUIZ

1. What types of valves are usually used in connection with emergency operations of landing gear, bomb-bay doors, and as system pressure dumps?

2. What type of check valve is used to retard the operation of flaps, landing gear, and similar gear in one direction?

3. What device acts as a junction box for the hydraulic lines, and is designed to save weight and space by eliminating many fittings?

4. What is the purpose of restrictors in hydraulic lines?

5- What device is used in unit systems where simultaneous operation of two or more actuating cylinders is desired?

6. Name the two types of flew equalizers.

7. The flow equalizer operates accurately under discharge port pressure differentials up to how much pressure per square inch?

8. What is the purpose of hydraulic control boosters in planes?

9. What device may be operated by the pilot from the cockpit in event fluid should become locked in the actuating cylinder of a hydraulic control booster and thus lock the control surface?

10. How are pressure gages used in hydraulic systems calibrated to read?

11. What auxiliary unit is installed in the line to the pressure gage for the purpose of dampening the oscillations and thus providing a steady reading and a protection for the gage?

12. In most installations using an electric motor for operation of a hydraulic pump, what takes the load off the pump when no units are being operated?

13. How much pressure is required'to hold up the arresting hook of the F4U-1?

14. What prevents the arresting hook of the F4U-1 from bouncing off the deck where a landing is made?

15. What devices have been added to later models of naval aircraft to prevclit loss of hydraulic fluid in case of a broken line?

16. Name two types of hydraulic fuses.

17. Name two advantages of solenoid-operated shut-offs or stop valves used in parts of the system remote from the cockpit.

18. In what three ways may electrically operated hydraulic units be controlled?

19. Of what, basically, does a solenoid consist?

20. What type of manifold block economizes in weight and space by incorporating both check valves and thermal relief valves?

CHAPTER 13
SHOCK STRUTS PURPOSE

A recent Army survey estimated that about 85 percent of structural-servicing time spent on aircraft is devoted to landing-gear units. This estimate leaves little doubt as to the tremendous importance of these units which include brakes, brake-control systems, shimmy dampers, and shock-absorber struts.

The" landing-gear struts we shall discuss in this chapter are of the combined hydraulic-pneumatic type, and the spring-loaded hydraulic type. Our coverage of these units will be only general, however, because there are so many different types of shock struts in current use.

Aircraft shock-absorber struts serve the fourfold purpose of absorbing and dissipating the landing shock on the compression stroke of the strut; absorbing and dissipating the jolts of taxiing; absorbing recoil shocks

which occur on the extension stroke of the shock strut on take-off, and acting as structural members to support the plane on the ground.

Both air and fluid are utilized in the struts to produce controlled resistance during aircraft take-offs and landings. The static weight is carried by air in the upper chambers, and this air, with the aid of gravity, serves to extend the strut so that it is in position to take the next shock load. The impact energy of the landing airplane is absorbed by metering of fluid. Shocks developed in taxiing are absorbed by air pressure.

Some types of struts use a steel compression spring in place of pressurized air to absorb the shock of taxiing and to bear the weight of the airplane in the static position.

CONSTRUCTION

In some installations, the strut takes compression loads only, while in others it takes bending and torsion loads as well. In the latter case, the strut is equipped with a sliding spline, or scissors, arrangement to prevent the piston from turning in its cylinder.

Almost all current types of aircraft use the type of strut in which hydraulic fluid and compressed air are combined to eliminate shock. The major internal parts of a shock strut are individually described below.

The telescoping members, known as the inner cylinder or piston and the outer cylinder, form an upper and lower chamber. The lower chamber is always filled with fluid, while the upper chamber contains compressed air. An orifice is placed between the two chambers through which fluid passes into the upper chamber during compression and returns during extension of the strut.

Most struts are equipped with a variable section pin known as a metering pin which automatically controls the area of the orifice at all points in the stroke of the strut.

The majority of struts—with the exception of tail struts—are equipped with an AXLE which is attached to the lower cylinder, to provide for tire and wheel installation. Those struts not thus equipped have provisions on the end of the lower cylinder for ready installation of the axle.

Suitable connections, or connecting members, allow proper attachment to the airplane for which each strut is designed.

A fitting consisting of a filler inlet and an air-pressure valve is located at the top end of each strut to provide a means of filling and inflating the unit.

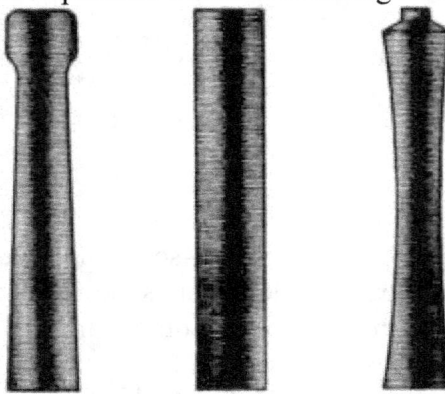

KNOB STRAIGHT CONCAVE TAPERED

Figure 60.—Four types of metering pint commonly used in struts.

The majority of struts are equipped with TORQUE ARMS attached to the upper and lower cylinders to maintain correct alinement of the wheels. Other struts—those without torque arms—have splined piston heads and cylinders which maintain correct wheel alinement.

Metering pins are always used in conjunction with either stationary orifices, orifices in flapper valves, or a combination of orifice and snubber valve. The four types of metering pins most commonly used in struts are illustrated in figure 60.

TYPES OF STRUTS

Nose gear, main gear, and tail gear struts may be divided into two categories— hard struts and soft struts. Planes with a tendency to be nose-heavy are equipped with hard main-gear struts, while heavier aircraft have soft main-gear struts. Nose-gear and tail-gear struts are usually of the soft type.

Struts are not distinguishable from the outside, and therefore are denned according to their operation. Soft struts compress rapidly on landing and extend slowly on

COMMUNICATIONS HOLES

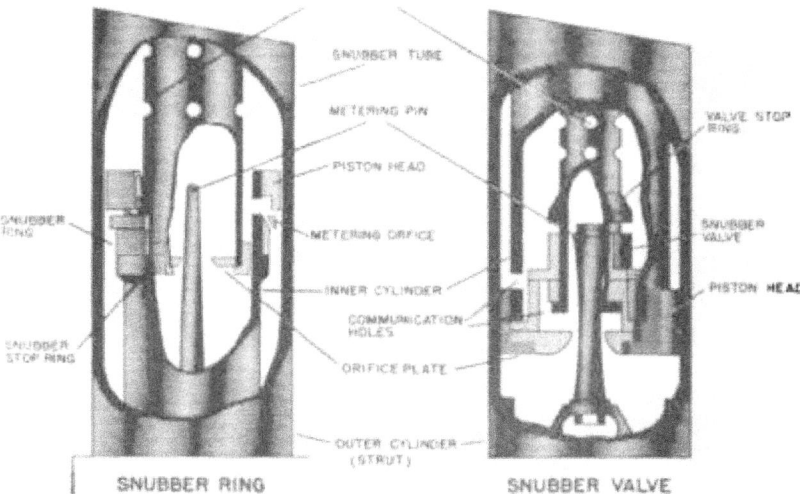

Figure 61.—Strut snubbing d.vic.$.

take-off. Hard struts compress slowly on landing and have rapid extension on the rebound.

The main gear (main gear and tail gear) on conventional-type planes is forward of the center of gravity.

This rapid or slow compression and extension of struts is accomplished by various snubbing devices, such as metering pins, flapper valves, snubber valves, orifices, and snubber rings. Typical example of snubbing devices are shown in figure 61.

Hard Struts

In most struts, the initial shock of landing- is absorbed by fluid metering through a metering-pin orifice.

In the illustration of the hard strut shown in the extended position in figure 62, notice that the knob is resting in the orifice. The fluid level is lower than in the compressed position, but the fluid still is above the orifice and the air is yet to be compressed. When the plane hits the deck and compresses the strut, fluid is metered comparatively slowly between the knob and the orifice, but when the knob gets past the orifice, fluid flows from the

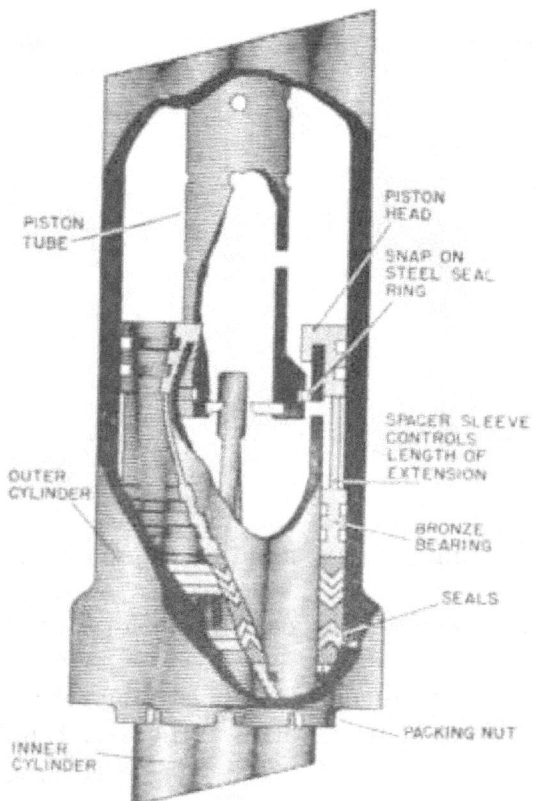

PISTON TUBE

PISTON HEAD

SNAP ON STEEL SEAL RING

SPACER SLEEVE CONTROLS LENGTH OF EXTENSION

OUTER CYLINDER

BRONZE BEARING

SEALS

PACKING NUT

INNER CYLINDER

Figure 62.-Principal parts of a typical hard strut (extended).

lower chamber, through the orifice, and into the upper chamber at a faster rate. As the fluid enters the upper chamber, it compresses the air until the entire load of the plane is carried on compressed air.

In the compressed position, the plane is actually riding on compressed air, and the "knob" of the metering pin is well above the orifice. When leaving the ground, compressed air and the weight of the wheel and inner cylinder will cause the inner cylinder to extend rapidly until the knob of the metering pin enters the orifice. This rapid extension of the strut is called the recoil stroke, and it can be readily seen that if this recoil stroke is not slowed, the resulting shock might damage the strut.

The rate of strut-extension is controlled by COMMUNICATION holes directly beneath the piston head.

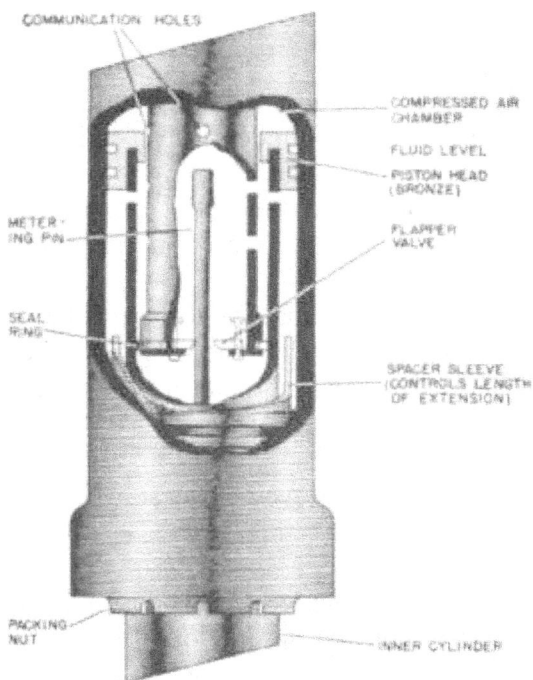

Figure 63.—Principal parts of a typical soft strut (compressed).

Soft Struts

The orifice in a soft strut is a moveable plate called a flapper valve. Metering-pin action of the orifice in soft struts is less effective than in hard struts.

The plane, in contacting the deck, compresses the strut and, as shown in the compression position of figure 63, fluid flow keeps the flapper valve open. Despite the open position of the flapper valve which allows greater fluid flow, only a certain amount of fluid can pass through the communication holes and past the set clearance between the flapper valve and its mounting plate. Consequently, the initial shock of landing is absorbed by fluid being metered through the flapper valve.

In the compressed position, the plane rides on compressed air with practically no fluid flow through the orifice. The flapper valve is therefore resting on its mounting. When the plane leaves the deck, fluid flow into the lower chamber keeps the flapper valve closed. The communication holes in the lower chamber are consequently closed, and fluid entering the lower chamber must pass through the orifice.

OLEO PNEUDRAULIC (AEROL) SHOCK STRUT

The action of the Aerol shock strut is easily understood. Although various models differ in design, most units operate upon the same principle.

The Aerol shock strut uses the scissors arrangement to prevent rotation of the piston within its cylinder. At the bottom of the strut may be seen the stub axle to which the wheel and brake assembly is fitted. Torsion links, or scissors, form a flexible connection between the strut cylinder and piston tube. Air is put into the compressed air chamber according to the amount necessary to obtain correct extension of the piston with relation to its cylinder. The oil chamber is filled with fluid when the piston is at the bottom of its stroke in the cylinder.

The highly polished piston tube is chrome-plated. Leak-age of fluid between the piston and cylinder is prevented by special packing held in compression by a packing retaining nut screwed into the strut cylinder. The point of the metering pin governs the amount of energy dissipated when the strut is compressed under the load of landing.

Aerol Shock Strut Operation

The compression stroke of the strut occurs when the impact load of the landing airplane is applied to the landing gear, and the piston telescopes into the cylinder. The speed of the compression stroke is determined by the rate at which fluid is displaced from the piston chamber. The tapered metering pin determines, in part, the rate of flow from the piston chamber. As the piston compresses, the air pressure and the tapered pin increasingly restrict the action, thereby requiring more energy to compress the strut.

During the extension stroke, no HIGH rebound exists because the flapper valve closes, requiring the fluid to be metered back through the fluid return holes in the snub-ber tube. The variable speed of the compression stroke is controlled by the tapered metering pin. The extended stroke is controlled by the fluid return holes in the snub-ber tube.

LANDING GEAR STRUTS

Some structural variations in nose gear, main gear and tail gear struts should be understood before we take up the details of servicing.

Main Landing-Gear Strut

Figure 64 shows a sectional view of a typical landing gear strut. This design is known as the hydraulic-pneumatic (air-oil) type. Hydraulic fluid is used for dissipation of the energy, and compressed air for carrying the static weight of the airplane and for cushioning the shock when taxiing.

In figure 64, the section of the strut labelled A is the

stub axle to which the wheel is fitted. Section B is fitted with a supporting member which also has extension brace attachment at C, at which point drag load bracing and retracting strut may be connected. Section D shows torsion links which prevent the wheel from rotating around the axis of the piston tube and at the same time allow the piston to telescope into the cylinder. Double links (torque arms, or scissors) may be installed when skis are used in place of wheels. Section E represents the compressed air chamber. The average pressure required to sustain the static weight of an airplane is approximately 500 pounds to the square inch.

The adjustment of struts is not gaged as is a tire, but

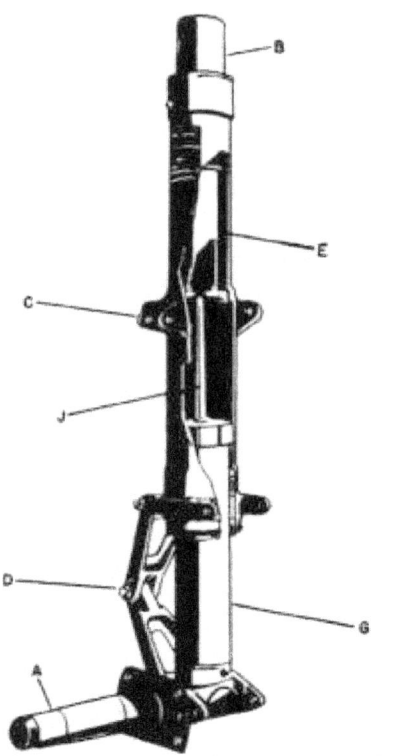

Figure 64.—Sectional view of a typical landing gear strut.

rather by the position of the piston. The average adjustment of the piston is approximately 20 percent extended. The required static pressure will diminish to a nominal pressure when the strut is fully extended. The oil chamber, normally filled with fluid when the piston is fully compressed, is shown as section E. Filling is made through a plug provided for the purpose, and so located as to furnish the proper compression ratio to the air chamber. Section G represents the piston tube, and the metering pin is shown as section J. This latter unit is of great importance. It is this pin which governs the amount of energy that is dissipated when the strut is compressed under load, such as in the landing of the airplane.

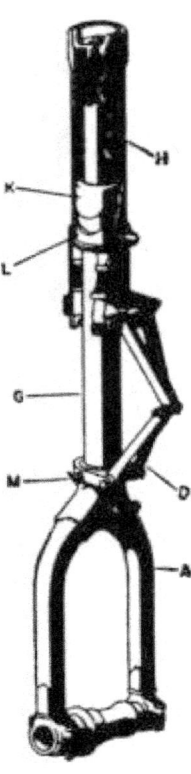

Figure 65.—Sectional view of a typical nose-gear strut.

Nose Gear Landing £rrut

In figure 65 is illustrated a sectional view of a typical nose gear landing strut. When landing an airplane, the nose strut functions, as far as air and fluid are concerned, much like the main landing-gear strut we have just studied. Additional parts, however, are added to achieve the special purpose of the nose struts.

As shown in figure 65, an upper locating cam labelled K, attached to the piston tube G and a mating lower locating cam L, attached to cylinder H, serve to line up the fork and axle assembly A in the straight-ahead position when the strut is fully extended. The nose wheel can then be retracted into the fuselage without interference. These cams also serve to keep the wheel in a straight-ahead position prior to landing when the strut is fully extended. To facilitate quick turning of the airplane when it is standing idle, a locking pin M is attached to the collar on which the lower torque arm D hinges.

Tail Gear Landing Strut

In figure 66 is illustrated a typical tail-wheel strut assembly consisting of a small strut wheel half-fork and supporting arm. The wheel is mounted on axle A. The half-fork (knuckle) N is mounted on roller bearings in the supporting housing 0, which permits the knuckle to caster and trail the airplane when taxiing. At the upper end of the knuckle assembly, a friction plate is fitted at P to offer rotating resistance to the knuckle N to reduce tail-wheel shimmy to a minimum. The section labelled Q is a typical tail-wheel shock strut, the operation of which is very similar to that of the main-gear shock strut.

SHOCK STRUT SERVICING

The length of life and satisfactory operation of the struts is largely dependent upon the maintenance and operating conditions of the other component parts of the landing-gear unit. Therefore, routine inspection of the

Figure 66.—Typical toil-wheel ttrut assembly.

wheels, tires, and all other units should be made at the time of the strut inspection.

It is advisable to check struts—at intervals dependent upon type of operation—to insure their correct extension- This is important in order to maintain the .shock-absorbing medium for taxiing and landing.

In daily visual checks, examine the air valve core for leaks by putting a small quantity of fluid on top of the valve. Air bubbles will form if the valve is leaking or is not properly seated. If leakage of the valve is observed, depress the valve core and allow it to snap back. This will permit the valve to reseat itself. If the leakage persists, the strut should be deflated and the valve core replaced with a new part.

Inspect the filler plug assembly for air leaks by placing a small quantity of fluid around the copper gasket. If leakage is apparent, the strut should be deflated, the filler plug removed, and a new gasket installed.

The struts should be inspected for fluid leaks around the packing. This failure is indicated by the presence of fluid passing the packing and bearing.

To check the fluid level, back off the filler plug slightly and allow air to escape slowly until the shock strut is fully compressed. Never remove the filler plug as

LONG AS AIR IS ESCAPING.

To service this unit, fill the shock strut through the filler plug hole with hydraulic fluid. After installing the filler plug with a new metal crush washer, charge with air to obtain a static load extension as specified on the servicing plate attached to each strut.

When filling the shock struts with air, the airplane should be rocked to prevent binding as the strut extends. Should the shock strut be over-inflated, remove small amounts of air by loosening the filler plug slightly and then quickly tightening.

The exposed reciprocating parts of shock struts are prone to pick up dirt and dust which is carried into the strut during operation, and deposited around the bearings to result, in many instances, in packing failure or scoring of piston walls.

Carrier-based aircraft are often exposed to salt spray and smoke which leaves a coating of salt and soot on the struts. These substances are carried into the cylinder during retraction and lodge in the packing, destroying the feather-edge. The result -is excessive leakage.

Inspect the entire outer portion of the strut assembly, especially in the locality of welds, lugs, and connections, EVERY 30 hours for any evidence of cracks or indication of structural failure.

In addition to daily and 30-hour inspections, inspect carefully all landing-gear attachment fittings every 120 hours. If necessary, remove the weight of the airplane from the landing gear in order to facilitate a thorough inspection.

QUIZ

1. What four purposes do aircraft shock-absorber struts serve?

2. What is the part in struts which automatically controls the area of the orifice at all points in the stroke of the strut?

3. What is the purpose of torque arms in struts?

4. What are the two main categories of struts?

5. Which of the two types of struts compress slowly on landing and extend rapidly on the rebound?

6. In what position of the hard strut is the plane actually riding on compressed air?

7. What controls the rate of strut-extension in the hard strut?

8. What in soft struts absorbs the initial shock of landing?

9. What in the Aerol shock strut prevents rotation of the piston within its cylinder?

10. What determines the speed of the compression stroke in Aerol shock struts?

11. In the typical hydraulic-pneumatic main landing-gear strut, compressed air is used for carrying the static weight of the airplane and for cushioning the shock when taxiing. For what is the hydraulic fluid used?

12. What on-the-spot remeay is recommended for leaky or improperly seated air valve cores in shock struts?

13. How often should the entire outer portion of the strut assembly be inspected for evidence of cracks or indication of structural failure?

14. How often should all landing-gear attachment fittings be in-pected?

15. Why should the airplane be rocked while the shock struts are being filled with air?

CHAPTER 14

BRAKE ACTUATING SYSTEMS PURPOSE

In previous discussions, we have considered the joint and individual operations of various units comprising the aircraft hydraulic system. We have seen how these elements are combined for the actuation of vital airplane mechanisms such as landing gear, bomb-bay doors, turrets, and wing flaps, and many others. Now we shall look into the composition and action of a separate and distinct hydraulic system that is at the same time an integral part of the vast and intricate hydraulic make-up of the airplane—the brake actuating system.

Our study of brake actuating systems will be like discussing the operation of a subsidiary firm in a far-flung corporation—the brake equipment, like the firm, functions independently, although it is part and parcel of the great main system.

An airplane brake actuating system is designed to

permit application of the brakes for landing, taxiing, and parking. Provision exists for applying either one or both brakes any desired amount by operation of the foot pedals.

TYPES AND USES

Brake actuating systems may be operated either mechanically, hydraulically, or pneumatically.

Mechanical-type brakes are now used only on older and lightweight airplanes. Cables, pulleys, and bell cranks are employed to connect the foot pedal to the brake shoe.

Hydraulic brake systems may be part of the main hydraulic setup or may be entirely independent systems. Independent, or master cylinder brake systems, are used at present on most of the comparatively light planes.

Larger aircraft employ a power boosted master cylinder brake system or a power brake control valve system. These two types form a part of the main hydraulic system. Power boosted master cylinder systems are now required on newer design carrier-based aircraft which are too heavy for master cylinder systems. This is due to the fact that enough braking action to hold the airplane against carrier roll can be obtained from power boosted master cylinders when the main hydraulic system is not pressurized—that is, when the engine is not running. Power brake control valves offer no braking action unless the main hydraulic system is pressurized.

Use of power brake control valve systems is also limited to aircraft having tricycle-type landing gear, since admission of system pressure to brakes because of failure of seals or metering mechanism would result in "nosing over" of aircraft not equipped with nose-wheels.

TYPICAL MASTER-CYLINDER BRAKE SYSTEM

The master cylinder is the energizing unit of the hydraulic brake system when the brake control is separated from the main hydraulic system. The purpose of the master cylinder—one of which is supplied for each

wheel brake—is to supply pressure for the brakes. This mechanism maintains the correct volume of fluid under all climatic conditions by compensating for the change in volume due to expansion or contraction. It also automatically replaces any fluid lost through leakage and prevents the entry of air into the system.

A master cylinder is essentially a manually-operated, single-acting reciprocating piston pump. It is usually mounted on the rudder pedal of the airplane and is controlled by a toe-pedal which is part of the rudder pedal assembly.

Most master cylinders have their own RESERVOIRS to make available an adequate supply of fluid to compensate for slight leaks in the connecting lines. As the reservoir is vented to the atmosphere to provide gravity feed to the pressure chamber, the correct fluid level must be maintained or air will be introduced into the system.

Mechanical linkages made up of rods and levers connect the brake pedal to the master cylinder. As the brake pedal is pushed down, the linkage causes the master cylinder piston to move. When foot-pressure is removed, the pedal is returned to the off position by the action of springs in the system. Although the brake pedals are part of the rudder pedal assembly, the brakes may be independently operated, or operated with the rudder.

Master cylinder brake system fluid lines may be either flexible or a combination of rigid and flexible tubing.

Brake actuating cylinders are located in the brake assembly. When pressure from the master cylinders is transmitted to them, they cause the brake shoe to press against the brake drum.

Some hydraulic brake systems include a parking brake mechanism with a control accessible to the pilot. The brakes are locked in the on position by depressing the brake pedals and then pulling the parking-brake lever. They are released by depressing the brake pedals. This action will either unload the ratchet-type parking

lock or build up sufficient pressure to unseat the parking valve, depending on the type of master cylinder. When the pedals are released, the master cylinder pistons return to the off position.

A compensating port is installed to prevent the brakes from being applied unintentionally by pressure of fluid expansion resulting from temperature increases. It also serves the purpose of insuring that the operating chamber is completely filled with fluid at all times. This compensating port provides a passage between the fluid reservoir in the master cylinder and the operating chamber whenever the brake pedal is released. The fluid is thus permitted to flow from the operating chamber to replenish fluid lost by leakage or contraction.

TYPICAL MASTER CYLINDER BRAKES SYSTEM OPERATION

The principles of operation of the various types of master cylinder brake systems are fundamentally the same, differences existing in structural details of the various units rather than in their function. Each brake pedal operates a separate master cylinder, allowing the pilot to manipulate either brake individually.

Brake fluid flows from the reservoir by gravity to the pressure chamber of the master cylinder. When force is applied to the brake pedals, the pistons are moved in the master cylinders, causing the fluid to flow through the lines to the actuating cylinders. Since the fluid is restricted to the system, its pressure rises with increased force on the pedals. On the return stroke, the various return springs, acting against their respective pistons, force the luid back to the master cylinder.

GOODYEAR MASTER CYLINDER

Of the four distinct types of master cylinders in general aircraft use, the Goodyear high-pressure model has been selected for the leadoff spot in our discussion of specific types because it offers the most easily understood features. The term high-pressure is applied to a cylinder

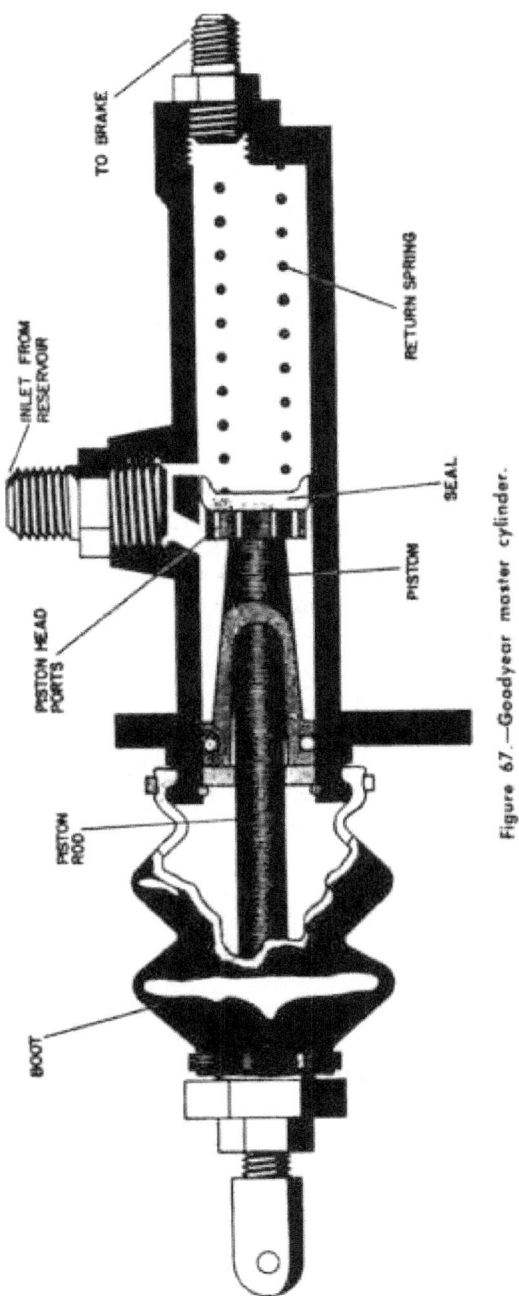

Figure 67.—Goodyear master cylinder.

when the diameter of the piston head does not exceed one and one-eighth inches. Master cylinders with piston heads of larger diameter are called low-pressure types.

In this unit, fluid is fed by gravity to the master cylinder from the external reservoir through the inlet and compensating ports as shown in figure 67. Application of the brake forces the piston forward, blocking the compensating port and building up pressure in the pressure chamber. This pressure is transmitted to the brake-actuating cylinder.

When the force is removed from the pedal, the piston is returned to the off position by action of the return spring, again clearing the compensating port. Any change in the volume of the fluid due to temperature changes while in the OFF position is compensated for by the passage of fluid through the compensating port. This insures against the possibility of locked or dragging brakes.

Locking of the brakes in the ON position for parking is accomplished by a ratchet-type lock built into the mechanical linkage between the master cylinder and the foot pedal. Any change in the volume of fluid due to expansion while the parking brake is on, is taken care of by a spring incorporated in the linkage. The brakes are unlocked by application of sufficient pressure on the brake pedals to unload the ratchet.

Fluid lost in front of the actuating cylinder piston through leaks in the connections, lines, or at the brakes, is automatically replaced by fluid passing through holes in the piston head and around the lip of the front piston seal when the piston makes the return stroke to the full off position. The front cup-type seal functions only as a seal during the forward stroke. The seal in the rear piston seals the rear end of the cylinder against external leakage at all times.

The flexible rubber boot protects the end of the cylinder from dust and dirt.

BENDIX MASTER CYLINDER

The Bendix master cylinder incorporates a reservoir, a pressure chamber, and a compensating chamber in the same unit. The reservoir is vented to the atmosphere through a small ball check in the filler cap, which prevents the loss of fluid when the plane is in inverted flight. Fluid enters the pressure chamber from the reservoir through the compensating port.

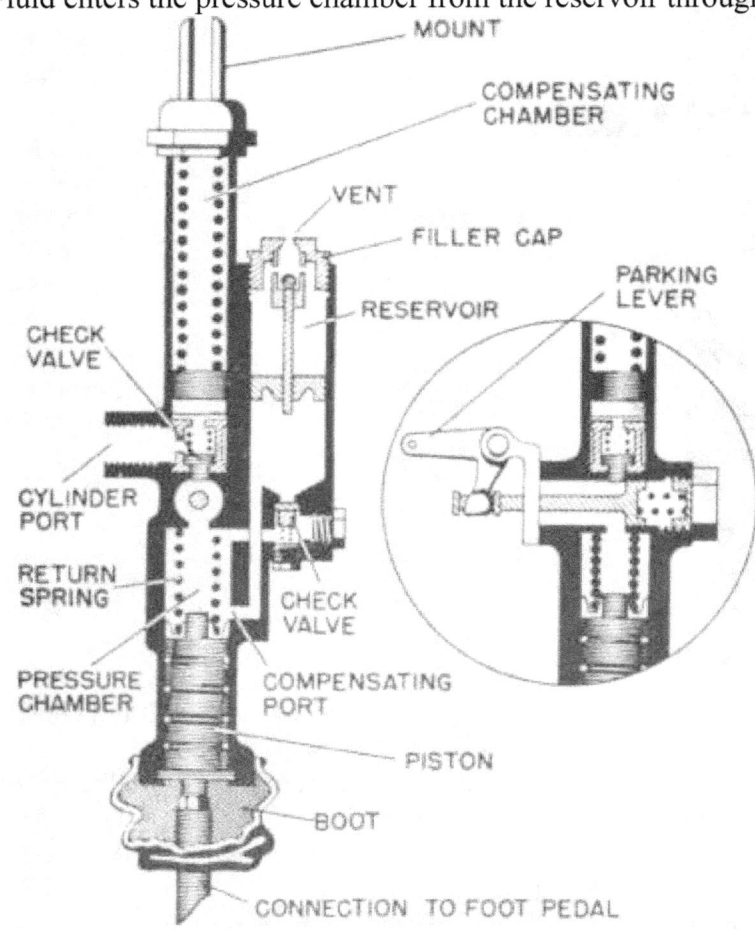

Figure 68.—Bendix master cylinder.

When pressure is applied to the foot pedal, the piston is moved upward, forcing fluid through a check valve and into the fluid line to the actuating cylinder. Before pressure can be built up in the system, the piston must move beyond the compensating port. Hence there is a cushioning effect which prevents violent locking of the brakes if they are applied too quickly.

As may be seen in figure 68, removal of foot pressure from the pedal allows the return spring to force the piston to the OFF position. Compensation for change in fluid volume due to temperature variations while the brakes are in the off position is accomplished by the compensating port which allows fluid to flow to or from the pressure chamber as required.

Pulling the parking brake lever while the brakes are being applied locks the brakes in the ON position for parking. By this action, fluid under pressure is trapped in the bottom of the compensating chamber and in the brake line. Pressure in the brake system is maintained by the partially compressed compensating spring. When there is a change of volume due to thermal effects while the brakes are parked, the compensating spring will either contract or expand, thus maintaining constant pressure.

WARNER MASTER CYLINDER

The Warner master cylinder houses a reservoir, compensating valve, and a reservoir check valve, as shown in figure 69.

Force applied to the brake pedal of this cylinder is carried by a mechanical linkage to the bottom actuating piston of the master cylinder. As the piston moves, the compensating valve closes, and pressure, which cannot escape through the closed compensating valve, is reproduced in the pressure chamber. The piston is moved to the off position by the return spring when pressure is removed from the pedal.

Increases or decreases in the volume of fluid when the

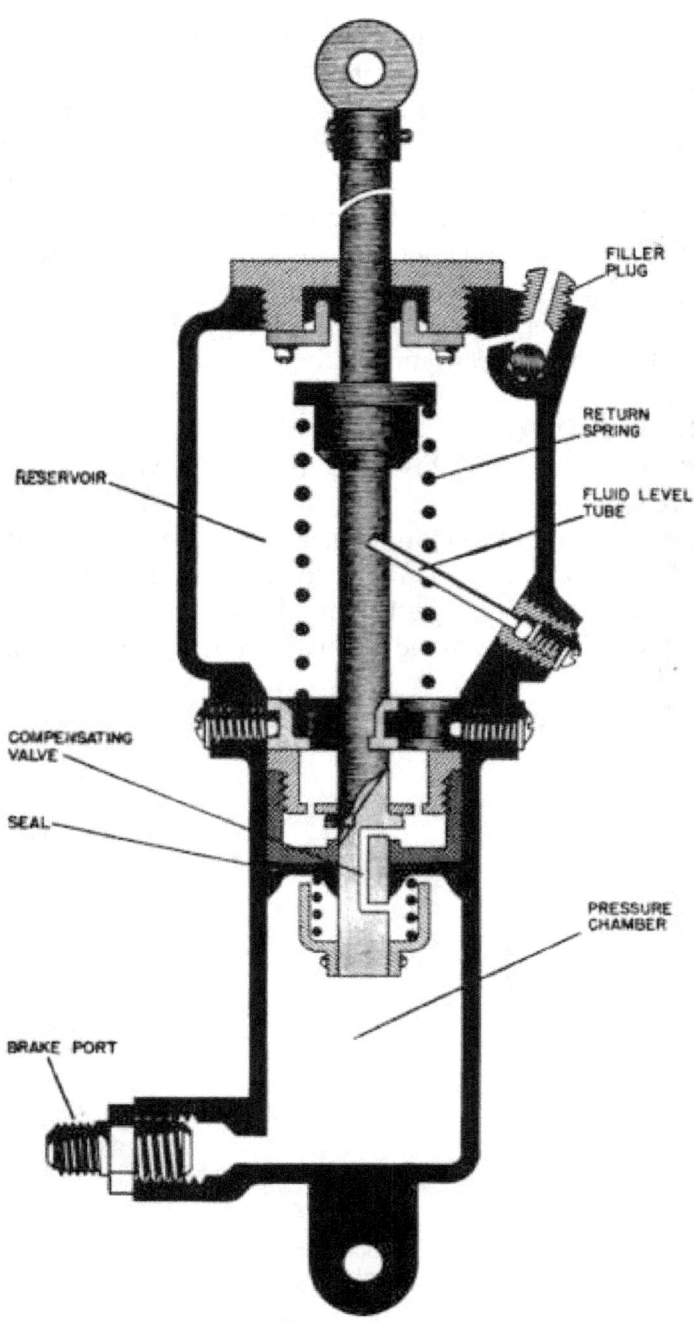

Figure 69 — Warner master cylinder.

brakes are OFF are compensated by the compensating valve which permits the flow of fluid between the reservoir and the pressure chamber. Locking of the brakes in the ON position is accomplished by engaging a ratchet which locks the piston in the ON position, thus compressing the parking spring. The purpose of the compensating valve is to allow the free flow of fluid from the reservoir to the brake line when the brakes are not applied.

TYPICAL POWER BRAKE SYSTEMS

Modern military aircraft are much heavier and have higher landing speeds than those of a few years ago, and consequently, larger wheels and brakes are required. Larger brakes mean either higher pressures or greater fluid displacement, or both, and for this reason, master-cylinder-type brakes would not be practicable on all types of aircraft. Fluid displacement is limited by the amount of force available for application to the pedal. The use of power brake systems is necessary to provide greater force.

Beginning at the main system, a line is tapped off from the main pressure line between the system accumulator and the first selector valve. The first unit in this line is a check valve which prevents loss of brake system pressure in case of main system failure.

The next unit is the accumulator, the main purposes of which are to store fluid under pressure for emergency operation, and to absorb surges when the power brake valve balances. The check valve works together with the accumulator in storing fluid under pressure. When the brakes are applied and pressure drops in the brake accumulator, more fluid enters from the main system and is trapped by the check valve. Not all power brake systems use a brake accumulator, in which case the main system accumulator takes up shock and surge from the power brake valves, but does not necessarily provide for emergency operation. In case of main system

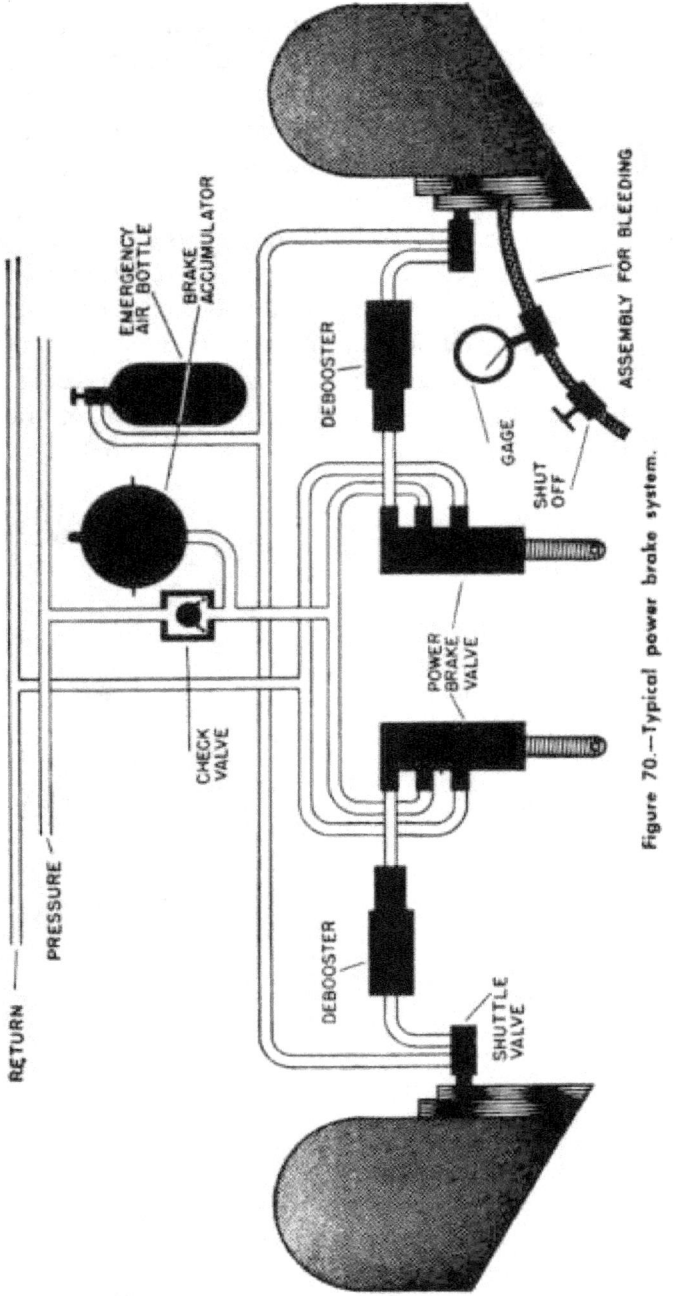

Figure 70.—Typical power brake system.

failure, emergency operation is provided by the hand pump or an auxiliary air-bottle used in conjunction with a shuttle valve which closes off the normal fluid entry port and opens the emergency fluid entry port. The shuttle valve is located on the brake assembly.

Next is the power brake valve, the purpose of which has been stated above.

De-boosters are sometimes used in power brake systems, especially where a high-pressure main system and low-pressure brakes—such as low-pressure multiple-disc, or expander-tube types—are used. The de-booster operates on the principle of difference in apea.

The main reasons for using the de-booster are to reduce pressure, obtain greater fluid displacement, faster return, and smoother operation. The de-booster is located in the brake line between the power brake valve and the brake assembly.

POWER BRAKE VALVES

The purpose of a power brake valve is to release and regulate pressure to the brakes.

Power brake valves vary in details of design and construction, but the same operating principles apply to all. All types have a pressure port through which fluid enters from the main hydraulic system; a return port through which fluid in the brake assembly can escape to the reservoir, and two brake ports through which fluid enters the brake line to the brake assemblies. When two valves are built into the same housing, a common pressure and return port is used.

Typical Power Brake Valves

When force is applied to the brake pedal, the piston in the brake valve is moved upward by means of mechanical linkage and the regulating spring. As the piston moves upward, the return poppet seats to close off the return. Further movement of the piston causes the pressure port check valve to open, allowing main system

RETURN SPRING

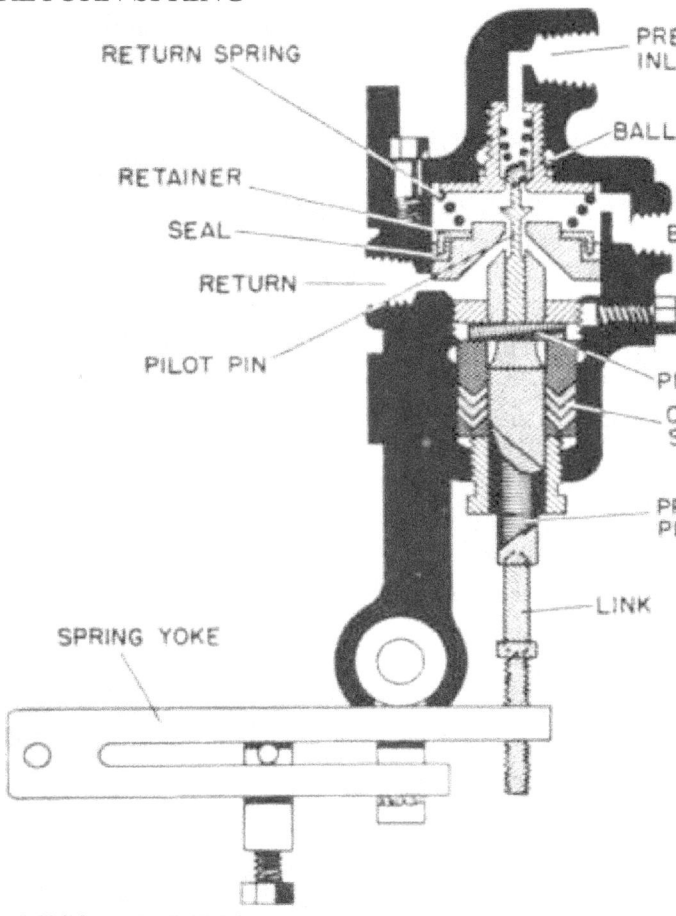

PRESSURE INLET
BALL CHECK
BRAKE PORT
PIN
CHEVRON SEALS
- PRIMARY PISTON

Figure 71.—Douglas power brake valve.

pressure to flow into the brake line. As the pressure increases in the brake assembly and line, the pressure also increases on the top side of the piston. When the total force on top of the piston is greater than the force applied at the brake pedal, the piston is forced downward against the regulating spring tension. This allows the pressure port check valve to seat, thus closing oft system pressure. In this position, the pressure and return ports are both closed and the power brake valve

is in the balanced position, meaning that a balance exists between the tension of the regulating spring and the fluid pressure on the top of the piston. This balancing action cuts system pressure down to brake pressure by closing off the pressure from the main system when the desired brake pressure is attained.

Spring tension and, therefore, brake pressure, are controlled by the amount of force applied at the brake pedal. As long as the power brake valve is in the balance position, fluid under pressure is trapped in the brake assembly and line.

All brake power valves have three chambers, named according to the parts to which they connect: Pressure chamber, brake chamber, and return chamber. Each is separated from the others by seals which may be either metallic, O-ring, or cup-type. If, due to leaks or any other cause, the pressure in the brake decreases, the pressure port will open to allow the pressure to build up and again equal the spring tension. If pressure in the brake increases, as it may because of thermal expansion, the piston will be forced down to overcome the regulating spring tension, and the return port will open to permit excessive pressures to escape. When either port continuously opens and closes, a power brake valve "cycle" occurs.

A leak from the pressure chamber to the brake chamber in the balanced position will cause the return port to open and close continuously. A leak from the brake chamber to the return chamber, in the balanced position, will cause the pressure port to open and close continuously. A leak from the pressure port, in the off position, will not affect the brakes, but will cause the main system regulator to cycle because the return port is always open when the brakes are in the OFF position.

Power brake valves have three positions—ON, OFF, and balanced. The balanced position has been explained. The off position occurs when no force is being applied to the brake pedal and the power brake is not

in operation. The pressure port is closed by the check valve and the return is open to allow free flow from the brake chamber to the reservoir. In the ON position, the brakes are being applied and the pressure port is open to the brake port.

The adjustments may be made on some power brake valves—the high-pressure adjustment which determines the highest pressure that may be applied to the brakes at full pedal throw, and the off adjustment which insures a positive free flow through the return port. All power brake valves have a return spring which returns the piston to the off position when the brake pedal is released.

A controlled amount of pressure is metered by the POWER BRAKE control valve from the main hydraulic system to the brake system, the amount of pressure admitted to the brake system depending upon the force applied to the foot pedal.

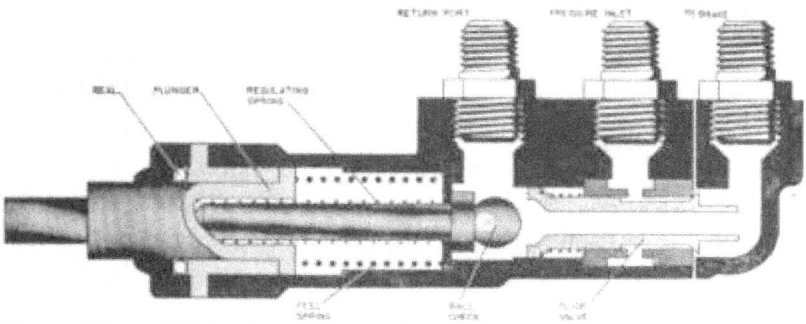

Figure 72.— Vickeri power brake valve.

Typical Power Brake Control Valves

A check valve and a piston containing a floating pin comprise the control valve assembly. The rod attached to the piston extends outside the housing and contacts one end of the spring yoke.

Fluid under pressure from the main system flows through the inport, past the check valve, and out of the

port leading to the wheel-brake assembly. Pressure acting on top of the piston produces a force which tends to move the piston downward, and when the pressure becomes great enough to bend the spring yoke, the piston and pin will move downward and allow the valve check to seat. When this occurs, no further pressure can reach the brake line.

When the pressure from the system builds up in the brake line until it exceeds the load of the ball on the spring, the balanced poppet valve reseats so that no more fluid is admitted to the brake line. Brake line pressure will be maintained as long as the brake pedal is depressed. Any leakage from the brake mechanism will be refilled by the balanced poppet valve unseating and admitting more fluid to the brake line.

To release the pressure at the brake, the pressure on the top of the rudder pedal is eased off, allowing the return spring to move the operating plunger out of the valve, and lessening the spring load against the ball. Pressure in the brake line will then unseat the ball, allowing fluid to return to the reservoir. When the operating plunger is fully retracted, or if the brakes are fully off, a minimum clearance of ^{2}-inch will exist between the ball and its seat which permits any brake-line pressure to flow through the hollow ball seat to the fluid reservoir.

BRAKE DE-BOOSTER VALVE

The brake de-booster valve consists essentially of a cylinder barrel which is fitted with a cylinder head on

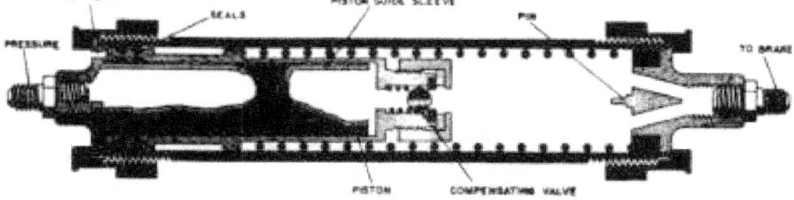

Figure 73.—Brake de-booster valve.

each end. These heads provide connections for the brake control valve and the brake lines.

Within the cylinder are a piston-guide sleeve and a spring-loaded piston, the latter dividing the cylinder into a small-volume, high-pressure chamber—connected to the brake control valve—and a larger volume, low-pressure chamber which is connected to the brake lines. A ball-type compensating valve is built into the center of the piston head.

Fluid under pressure from the brake valve enters the pressure port of the de-booster cylinder. The pressure acts upon the small area of the piston, forcing it against the fluid in the line to the brakes. The resultant to the brake is reduced at a constant ratio. This reduction is acquired by providing a greater piston area on the brake side of the piston for the force to act upon. When the pressure is released in the de-booster, the brake pressure and spring return the piston to its original position. The purpose of the ball-check in the piston is to relieve thermal expansion and facilitate bleeding. The ball-check also provides fluid to the brake line in the event of leakage in the line from the de-booster cylinder to the brake. This unit is automatically bled during the bleeding of the brake.

COMBINATION POWER BRAKE VALVE AND MASTER CYLINDER

The power-brake side of the valve functions exactly as the power brake valves described earlier in this chapter. If the power brake system should fail, the master cylinder operates exactly the same way as the master cylinder previously discussed.

As shown in figure 74, fluid is free to flow between the brake and the return when the power brake valve is in the OFF position. When the pedal is depressed and the clearance is taken up, the return closes. Further travel of the pedal opens the brake port and the pressure port and compresses, or puts tension upon, the regulating

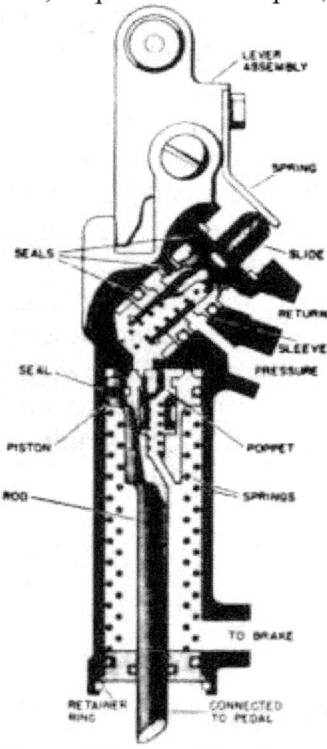

Figure 74.-Combination power brake valve and matter cylinder.

spring. When the brake is released, a return spring returns the linkage, and another return spring closes the brake port and the pressure port and opens the return. Fluid is then free once again to flow between the brake and the return.

The following points should clarify the operation of master cylinders. In the OFF position, fluid is free to flow between the brake and the return. The first thing occurring when the pedal is depressed and the clearance taken up is the closing of the return, or compensating, port. Further travel of the pedal creates flow to the brake actuating cylinder where the flow meets resistance and creates pressure which is controlled by the force of the pilot's foot on the pedal.

When the brake is released, a return spring returns the linkage and opens the return port, and fluid is again free to flow between the brake and return.

The Aviation Structural Mechanic will notice that the operation of brake control valves has been generalized, and that a comparison of operating principles may be applied to practically any hydraulic unit of a similar nature and designed for similar purposes.

INSPECTION AND MAINTENANCE OF BRAKE SYSTEMS

The operation of airplane brake systems must be faultless at all times. To assure this perfect performance, inspections must be conducted at regular intervals, and maintenance work which might be indicated as a result of such inspections should be performed promptly and carefully.

All lines are inspected with the system under full operating pressure when checking for leaks. Loose fittings must be tightened with the pressure OFF. All flexible lines should be carefully checked for swelling, cracking, and soft spots, and replaced if evidence of deterioration is noted. If leaks are found at the master cylinder or actuating cylinder, packing nuts may be tightened. Care should be exercised in tightening packing nuts to avoid binding of the rods.

Air in the system is indicated by "spongy" action of the foot pedals. If air is found in the system, it should be removed by bleeding as described below.

If the general operation of the system is unsatisfactory, some unit or other may require repair or replacement. Check all linkages carefully before removing any unit. Scored cylinders, bent piston rods, or other unserviceable conditions are sufficient justification for replacement. The replacement of cups, seals, or gaskets may constitute a satisfactory repair.

The proper fluid level should be maintained in the master cylinder reservoir at all times to prevent brake failure or the introduction of air into the system.

General Bleeding Procedure

Wherever hydraulic brake lines are disconnected, air will be admitted into the system. Insufficient fluid in the reservoir of a master cylinder system will also admit air. Naturally, air entering the system must be removed, and such removal is accomplished by bleeding.

The bleeding of individual systems will present individual problems. Therefore, the following methods of bleeding are merely general recommendations, and consultation of specific maintenance manuals is advised before bleeding of a particular system is undertaken.

Master cylinder hydraulic brake systems may be bled by pumping fluid through the system, using the master cylinder as a pump. In this method, the free end of a filler-can hose is attached to the reservoir filler-plug opening. The can is then partially filled with the same hydraulic fluid prescribed for use in the master cylinder. A hose is attached to the brake actuating cylinder bleeder-valve, and the free end submerged in hydraulic fluid in a glass jar, after which the brake pedal is manipulated back and forth, making rapid-downstrokes and retarding the return. During this phase of the operation, the bleeder valve is kept open during the down-stroke and closed before the up-stroke. Operation of the pedal is continued until air bubbles no longer appear in the glass jar. This may require the pumping of one or more pints of fluid through the system. The bleeder valve is then tightly closed and the hose is removed upon completion of the bleeding procedure. The fluid level should be checked and brought to the proper point.

Bleeding of master cylinder brake systems may be accomplished by forcing fluid into the system with compressed air through the actuating cylinder bleeder-valve. In this operation, the reservoir plug is removed, a nipple is inserted, and a hose is attached with the free end leading into a clean can. The actuating cylinder bleeder port is opened and the fluid line from the container is attached. (This fluid line should incorporate a shut-off valve and a pressure gage). The fluid in the filler container should be put under from 6 to 10 p. s. i. pressure by adding compressed air. With the brakes released, the shut-off valve in the line leading to the brake actuating cylinder should be slowly opened, allowing fluid to flow into the cylinder. Care should be taken to prevent pressure in the actuating cylinder from exceeding the minimum necessary to create fluid flow through the system. Allow the fluid to flow back (by gravity) through the pressure tank to remove air bubbles from the brake. When the fluid has run out of the hose from the reservoir filler-plug, the shut-off valve should be closed. The bleeder valve at the actuating cylinder may then be closed, and the filling equipment removed.

Bleeding of power brake control valve systems is accomplished by metering fluid through the power brake control valve by operation of the foot pedals. Before the bleeding operation begins, the brake accumulator must be charged. A bleeder hose is attached to the actuating cylinder bleeder valve and the free end of the bleeder hose is attached to a clean glass jar and submerged in fluid. The bleeder valve is then opened, and fluid is metered through the power brake control valve by operating the foot pedals until no bubbles appear in the bleeder jar. The bleeder valve is then tightly closed, the bleeder hose disconnected, and the dust cap replaced.

POWER BRAKE VALVE ADJUSTMENT

Adequate pressure-gage equipment, and an understanding of the operation of the complete hydraulic system, are essential when checking the brake system for service troubles. Always remember that pressure is dependent primarily upon the ability of the hydraulic system pump to deliver the hydraulic fluid at the rate required by the hydraulic system. Make certain that the pump is operating properly before attempting to change the adjustment of the brake control valve or to check it for faulty operation.

External-Spring-Type

Two adjustments may be made on the external-spring-type brake control valve—the off adjustment and the high-pressure adjustment. The OFF adjustment is always made first, and is accomplished by attaching a gage to the bleeder valve on the brake-adjusting cylinder and opening the bleeder valve. The piston is turned inward until the gage registers 10 to 15 p. s. i., after which the pedals are depressed halfway and released.

The 15 p. s. i. pressure indicates that the floating pilot pin is seated on the piston and that the return is therefore closed. The piston is then slowly backed off until the gage shows zero p. s. i., whereupon the screw is backed off one-fourth turn and locked. The zero pressure reading indicates that the pilot pin is unseated and the return consequently open. The additional one-fourth turn insures that the pilot pin is unseated when the valve is in the OFF position.

The high-pressure adjustment is made with full operating pressure in the system. A high-pressure gage is attached to the bleeder valve and the valve is opened. The high-pressure adjustment fulcrum is placed approximately 2 inches from the point of connection to the brake linkage, and the gage-reading is noted after the foot pedals have been fully depressed. If the pressure is too low, the fulcrum is moved toward the low-pressure adjusting screw, and if too high, the fulcrum is moved in the opposite direction. Bear in mind that applied spring tension is being increased or decreased in this operation. The adjustment is continued until the pressure is within the correct limits for the unit, after which the bleeder valve is closed, the gage removed, and the dust cap replaced. Do not remove the fulcrum when it is under load.

Internal Spring-Type

Adjustment of the internal spring-type brake valve is accomplished by varying the stroke

of the push rod. A

high-pressure gage is attached to the bleeder valve, and the valve opened. With full operating pressure in the system,' the foot pedal is fully depressed and the pressure-gage reading noted. If the pressure is too low, regulating-spring tension must be increased. If the pressure is too high, regulating spring-tension must be decreased. When the pressure is within correct limits for the unit, close the bleeder valve, remove the gage, and replace the dust cap.

QUIZ

1. How may brake actuating systems be operated?

2. Why is use of power brake control valve systems limited to aircraft having tricycle-type landing gear?

3. What is the purpose of the master cylinder supplied with each wheel?

4. In the Goodyear master cylinder, what insures against the possibility of locked or dragging brakes caused by a change in the volume of fluid as a result of change in temperature?

5. When a plane equipped with Bendix master cylinders is in inverted flight, what prevents the loss of hydraulic fluid?

6. How is locking of brakes in the ON position accomplished in the Warner master cylinder?

7. Upon what principle does the de-boosters operate?

8. For what reasons are de-boosters used?

9. What is the purpose of a power brake valve?

10. All brake power valves have three chambers, named according to the parts to which they connect. What are they?

11. What three positions do power brake valves have?

12. Name the parts comprising a typical power brake control valve assembly.

13. How much pressure should be maintained when checking lines of brake systems for leaks?

14. How may air in brake systems be detected?

15. What three methods are utilized to force hydraulic fluid through brake systems and thus accomplish bleeding of the system?

16. What two adjustments may be made on the external-spring-type brake control valve?

17. Under what system pressure is the high-pressure adjustment made?

18. How is the adjustment of internal spring-type brake control valves made?

CHAPTER 15

HYDRAULIC BRAKES PURPOSE

The moment an airplane lands, the hydraulic brakes are needed. Airplane brakes are designed to hold the plane on deck at full throttle; to stop the plane after hitting the deck; and to aid the plane in taxiing.

To aid in preventing serious accidents, such as ground looping, the Aviation Structural Mechanic must have a thorough understanding of brake operation and correct servicing procedures. This, then, will be the aim of our study in this chapter.

The important maintenance problems of hydraulic brakes are shoe-clearance and bleeding. We shall consider in this study the distinct types of hydraulic brakes used on aircraft.

BENDIX SHOE BRAKES

The INTERNAL EXPANDING SHOE BRAKE (Bendix) is

similar to those used on automobiles and trucks, and is employed chiefly on single-engine planes. These units are composed of a circular fiber brake shoe mounted on a metal frame. When the brake pedal is depressed, this shoe is forced against the inner surface of a circular metal drum.

Internal expanding shoe brakes are divided into two classifications— two-shoe servo and single-shoe servo brakes. The servo units are the actuating cylinders which change tha fluid pressure into mechanical motion.

The servo action is effective in single-servo brakes for one direction of wheel rotation only. Because of this, the brakes are not interchangeable between the right and left wheels of an airplane. The brakes are marked with the direction of rotation of the wheels with which they are designed to be used, such markings indicating the direction of rotation from the wheel side of the brake.

Single-servo two-shoe brakes have but one piston in the actuating cylinder, the piston acting to push one end of the brake shoe against the wheel-drum. This unit consists of a bracket, called the torque spider, used to secure the assembly to the landing-gear brake flange, a two-piece shoe, and an actuating cylinder.

The brake shoe attached to the piston rod is called the primary shoe, and the other part is termed the secondary shoe. Fluid pressure applied to the actuating cylinder causes the primary shoe to expand against the wheel drum as the piston is extended in the cylinder. Rotation of the wheel drum then causes the shoe to expand further and applies the secondary shoe, increasing brake force. The tendency of the drum to "wrap" the shoe onto the drum when the wheel is rotating is called energizing action and actually aids in applying the brake.

When fluid pressure in the actuating cylinder is released, a coil spring connected between the two shoes draws them away from the drum and releases the brakes, forcing fluid from the actuating cylinder back into the brake control valve.

The duo-servo brake, shown in figure 75, differs from the single-servo in that the end of the brake-shoe assembly, instead of being anchored to the torque spider by a stationary pin, is connected to a second piston in the actuating cylinder. When hydraulic pressure is admitted to the actuating cylinder, the application of force

is transmitted in opposite directions, through connecting-linkage, to the two ends of the shoe assembly.

The duo-servo brake is desirable because its braking action is effective in either direction of wheel rotation. It is "reversible" because the same brake assembly may be used on either right or left wheel.

SEAL

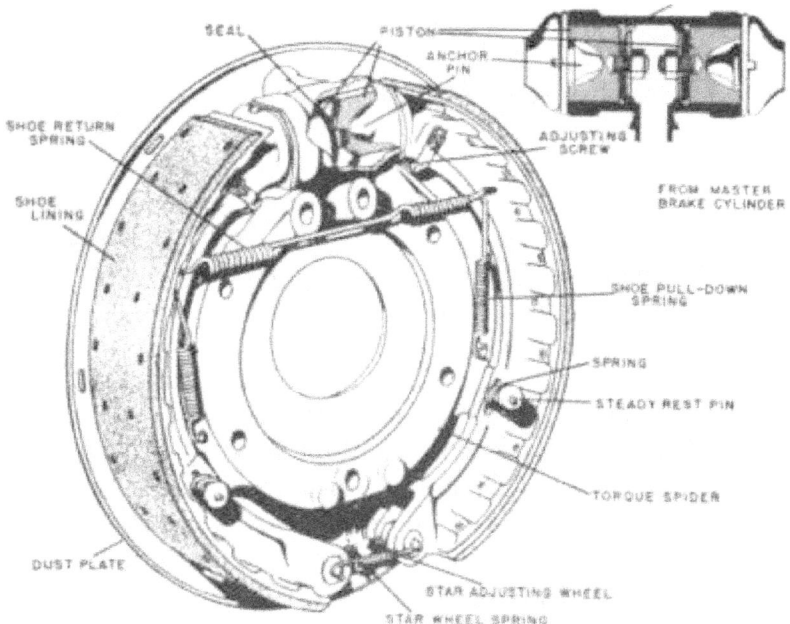

Figure 75.—Double-acting duo-Mrvo (two shoe) hydraulic brake.

Servo Brake Maintenance and Adjustment

On two-shoe duo-servo brake units, there are three adjusting screws, accessible through the dust cover, which may be operated with a screw driver. When making adjustments, the mechanic should use four feeler gages of proper thickness to insure uniform clearance all around the shoe. The clearance is generally between 0.008-inch and 0.012-inch, but may differ on different planes under varying operating conditions.

Adjustment Of servo brakes should proceed according to the following steps.

Jack up the wheel and check the wheel bearing-nut for track and fit by tightening the nut until the wheel drags. Loosen the nut one-eighth turn at a time until the wheel tracks freely, then turn off to the next keyway. Shake the wheel from the top to make certain it is not loose on the axle, then check the rotation by spinning the wheel. Be sure that the wheel bearing-nut has been keyed following the adjustment.

Now turn the star-wheel adjusting nut, located at the bottom of the brake, until the wheel rotates freely. Turn the eccentric adjusting screws, located at the top of the brake, until the wheel is locked. Then turn these screws off until feeler gages show 0.010-inch clearance, and lock the eccentric lock nut.

After this has been done, rotate the wheel and apply the brake several times. Tighten the star-wheel adjusting screw until the brake drags, then loosen the screw until the feeler gage shows .010-inch clearance. Rotate the wheel and apply the brake to check the adjustment.

Servo brakes must be individually bled because the brake lines are single-static—that is, there is no circulation in them. Spongy action of the brake pedals, along with excessive pedal throw, indicates that air is trapped in the brake cylinders or lines.

The procedure for bleeding brakes is as follows:

Remove the bleeder-fitting dust cap and attach an extension line to the bleeder fitting. Submerge the free end of the bleeder extension line in a bottle of the proper-type fluid. Then loosen the bleeder-valve hex nut one turn and slowly operate the brake pedal until air-free fluid emerges from the brake. In this step, the hex nut should be closed during each return stroke of the pedal if the master cylinder is supplying brake pressure. Close the bleeder-valve hex nut, remove the bleeder extension line, and replace the dust cap. Finally, test the brake for proper function.

A highly important precaution to observe when installing single-servo units is to ascertain that the brake is fitted on the correct right or left axle.

HAYES REVERSIBLE BRAKE

The Hayes reversible brake, while not termed a duo-servo unit, has two actuating cylinders. However, the Hayes brake has but one shoe. The assembly is interchangeable, but when transferred from one side of the plane to the other, or when a new assembly is installed,

Table 9.— Hayes reversible brake service trouble and

the mechanic must make certain that the red spring is forward and the black spring is aft.

Although operating conditions may require a variation in clearance adjustments, the recommended clearance on the Hayes brake is 0.008-inch to 0.010-inch.

To adjust the Hayes brake, jack up the wheel and adjust the brake after checking the wheel bearing nut for track and fit. Check the adjustment with four feeler gages, then tighten all adjusting screws, in rotation, to a snug fit. Loosen each adjusting screw, in rotation, one-half turn at a time until the correct clearance is obtained, and check clearances and operation by several applications of the brakes as well as with the feeler gages.

EXPANDER-TUBE BRAKE

The expander-tube brake has three main parts—the brake frame, the expander tube, and brake blocks. Braking action in this unit depends upon fluid under pressure expanding in the tube. This expansion, in turn, forces the brake blocks against the brake drums. The expander-tube brake may be used with any conventional hydraulic brake system.

Two types of expander-tube brakes are in current use. The single style has one row of brake blocks around the circumference, while the duplex style has two rows of blocks. These blocks are made of special material. Their total number corresponds to the number of inches of brake diameter. Thus, a 20-inch brake would have 20 blocks. Notches in the block sides engage with bosses on the brake frame so that they will not rotate around the frame. Flap springs fit into slots in both the blocks and frame, holding the blocks against the expander tube, and preventing them from dragging when the brake is released.

When the pilot applies the brake pedal, the pressure sends fluid into the expander tube. Since this tube is restrained from moving inward or to the sides, its fluid

forces it to expand outward and apply the brake blocks against the brake drum. Upon release of pressure within the expander tube, the springs in the ends of the block force them against the expander tube and eject the fluid from the tube; the elastic tube itself, slightly smaller than the brake drum, assists in this action.

Each block is independent in action. Since no build-up of servo action exists, no tendency of the brakes to "grab" is present.

Adjustment and Maintenance

The expander-tube brake operates at low pressures, usually from 200 p. s. i. to 300 p. s. i. Because of this the brakes should never be applied with the wheel removed.

These units are self-adjusting. If the blocks appear to be wearing unevenly, jack up the wheel and check the wheel bearing for track and fit.

The best clearance for this brake is between 0.025-inch and 0.030-inch, but the clearance should not exceed

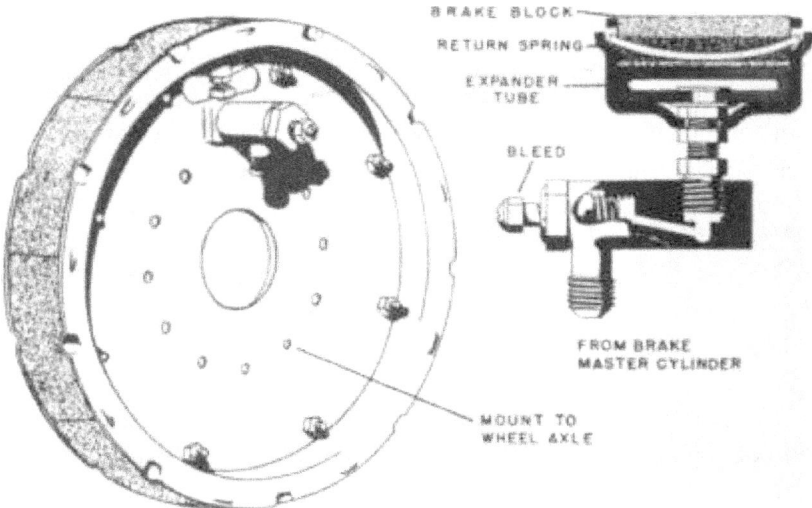

Figure 76.—Expander-tube hydraulic brake assembl.

0.075-inch or fall below 0.015-inch. If the clearance is less than the minimum, remove the wheel and file oversized blocks to the correct dimensions. If all blocks have worn to greater than 0.075-inch, remove the wheel and replace the tube and blocks.

Expander-tube brake assemblies are interchangeable, but on installation, make certain that the bleeder fitting is within 15 degrees of the top.

Shoe wear in this brake is compensated for by a natural expanding action of the tube, brought about when fluid causes the rubber in the expander to grow in size. However, as has been stated, if the clearance rises over 0.075-inch, new blocks and a new tube should be installed.

Bleeding procedure on the expander-tube brake is similar to that outlined for the servo units.

Table 10.— Expander-tube service trouble and remedy chart
continued

MULTIPLE-DISC BRAKE

The multiple-disc brake is very effective for stopping heavy aircraft. This unit consists of alternating stationary steel discs which are keyed to anchor brackets and rotating bronze-plated steel discs which, in turn, are keyed to the wheel. All discs are free to move back and forth in a line perpendicular to their plane of rotation. That is, the discs are free to slide endwise on the bracket, but cannot rotate with the wheel. These discs are known as rotor plates.

Disc travel is limited on the inboard side by a circular cup seal and piston, and on the outboard side by a circular flange-locking ring threaded onto the anchor-plate assembly. The braking action is produced by contact between the stationary discs and the rotating bronze-plated steel discs. Hydraulic pressure against the circular piston makes the contact.

With the brake in the released position, a clearance, adjusted by the plate retaining nut, is maintained between the plates. An annular groove running completely around the circumference of the anchor bracket serves as an actuating cylinder.

Adjustment and Maintenance

The following rule will enable the mechanic to obtain proper clearances in the multiple-disc brakes.

Allow 0.003-inch clearance per disc. Count the total number of discs, multiply by 0.003-inch, turn the adjusting nut down snugly, and back off one turn for each 0.060-inch clearance required. Back off to the next locking screw lug, and lock securely in place.

If the clearance desired, for example, is 0.075-inch, then the adjusting nut is first tightened and then backed off one and one-fourth turns, and further, if necessary, to the nearest locking screw lug. One turn is equal to 0.060-inch, so one-fourth turn will equal 0.015-inch, and one and one-fourth turn, therefore, equals 0.075-inch, the total clearance required.

The amount of braking surface, and thus the amount of braking action for any one pressure, can be varied by altering the number of plates. The amount of braking surface may also be changed by increasing or decreasing the diameter of the plates. Therefore, on heavy aircraft, the brakes normally would consist of a large number of large-diameter plates. On a smaller plane, both the number of plates and their size would be smaller.

To adjust the multiple-disc brake, remove the wheel and then the cotter key from the flange lock screw. Adjust the brakes with the adjusting nut, checking clearances with the use of several feeler gages inserted between the adjusting rings and the first steel disc. In the absence of feeler gages, the adjusting ring may be

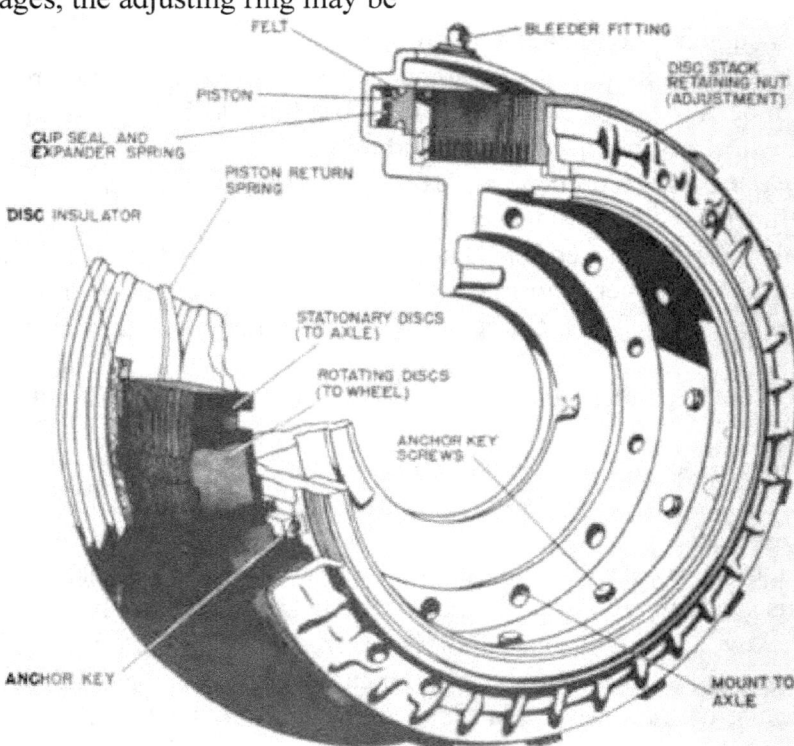

Figure 77.—Multiple-disc hydraulic brake assembly.

brought up fully tight and then backed off sufficiently to give the proper clearance. When the adjustment is completed, lock the adjusting ring with the lock screw and cotter key.

In bleeding this type of brake, observe the same procedure as with the servo units.

SINGLE-DISC BRAKE

Single-disc brakes operate both mechanically and hy-draulically. There are two types of hydraulically operated single-disc brakes, the fixed-disc type and the floating-disc type. Both models may be installed on either the right or left wheel.

In the fixed-disc type, the disc is a part of the rotating wheel and the pistons push the brake linings against the disc from both sides. The brake assembly is bolted to the torque plate, or wheel-axle mounting flange, through bushings which allow a slight sidewise movement of the brake assembly when the brake is applied. Thus, equal pressure is obtained on each side of the disc.

The floating-disc type is one in which the disc is keyed to the wheel in a manner comparable to the rotating discs of the multiple-disc brake. When the brake is applied, the piston moves outward, pushing the disc against the stationary brake lining on the outboard side of the wheel. This action equalizes pressure on both sides of the disc, as the brakes are applied.

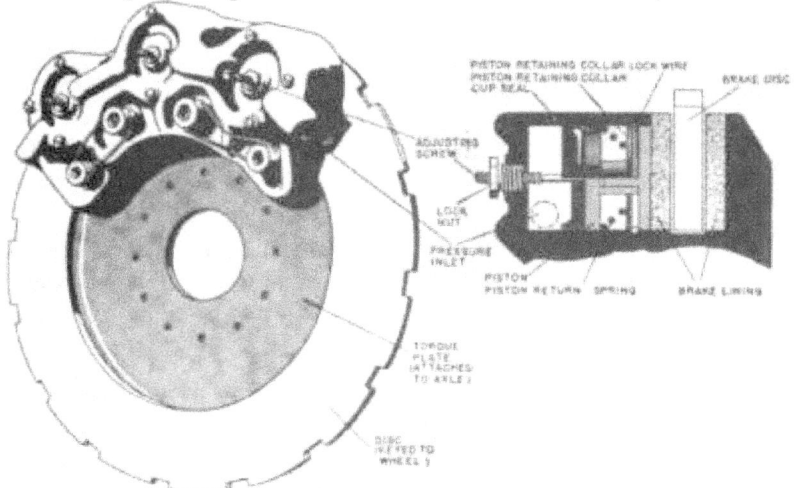

Figure 78.—Single-disc hydraulic brake assembly.

Adjustments are made on clearances until brake linings are worn to about 0.160-inch in thickness, when they must be replaced. Adjustment is accomplished by loosening the adjusting-screw lock nut and turning the adjusting screw until proper clearance is obtained. A feeler gage is inserted between the disc and the inboard brake-lining segment. Clearance is to be checked at this point on all brakes of this type.

QUIZ

1. What are the important maintenance problems of hydraulic brakes?

2. Name the two classes of internal expanding shoe brakes.

3. Why are single-servo brakes not interchangeable between right and left wheels of an airplane?

4. Why must servo brakes be individually bled?

5. What are the three probable causes for servo brakes grabbing?

6. How many shoes has the Hayes reversible brakes?

7. What is the recommended clearance on the Hayes brake?

8. What are the three parts of the expander brake?

9. At what pressure does the expander-tube brake operate?

10. What remedy should be applied to dry wet linings in planes equipped with expander-tube brakes?

11. To obtain proper clearances in the multiple-disc brakes, how much clearance is allowed per disc?

12. Name the two types of single-disk brakes.

CHAPTER 16 SHIMMY DAMPERS

If you have ever given an office swivel-chair a push, and seen it tip over because its casters shimmied, then you can appreciate what would occur if the nose-wheel of an airplane were rigged in the manner of the office chair.

The nose-wheel of a plane is protected, however, from such occurrences by the installation of a device called the shimmy damper, whose purpose is to prevent the dangerous shimmy, or oscillation, otherwise prevalent in tricycle landing gear. Certain types of shimmy dampers also act as nose-gear steering mechanisms, for use in taxiing or towing the aircraft. Still other types are designed to allow a slow flow of fluid from one chamber to another to permit turning of the nose-wheel.

PISTON-TYPE SHIMMY DAMPER

Let's look first at the piston-type shimmy damper, since that model is the simplest and the most common of several variations of this unit.

The piston-type shimmy damper consists of two cylinders attached to the outer cylinder of the nose-wheel

shock strut, with the piston rods connected to the inner cylinder of the shock strut. As shown in figure 79, a fluid line connects an accumulator to corresponding ends of each cylinder.

A comparison with the illustrations in the chapter on shock struts will disclose that these shimmy damper cylinders have snubber valves very similar in appearance to those attached to struts. The shimmy damper snubber valve operates in practically the same way as a strut

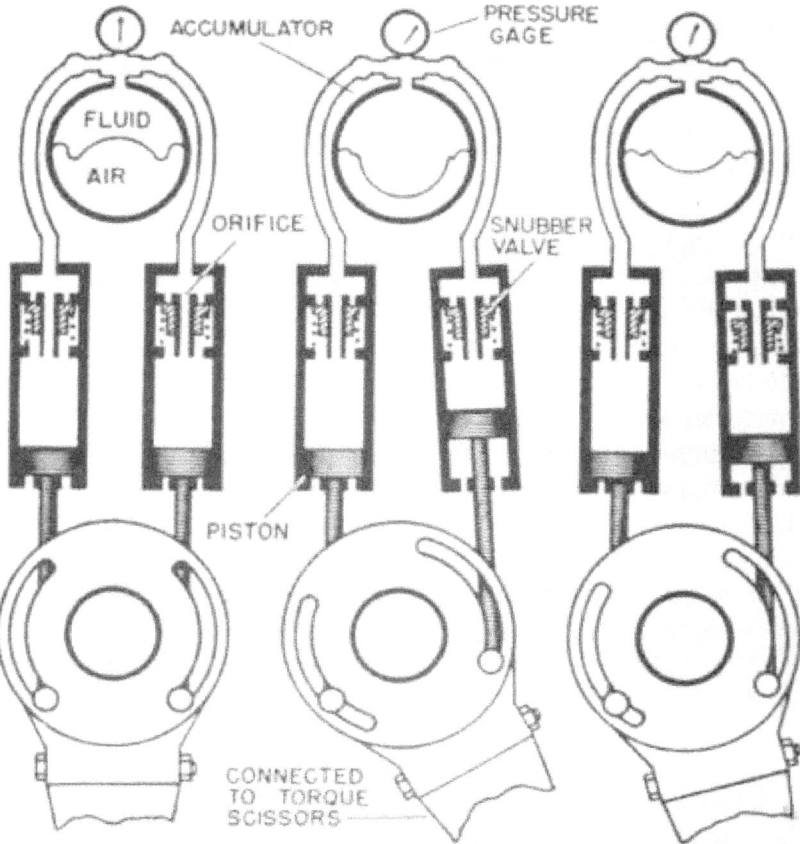

NEUTRAL BEING COMPRESSED RETURNING
Figure 79.—Pitton-type shimmy damper.

snubber valve, allowing free fluid flow when returning to neutral, and restricting flow when the wheel tends to swing out of line. The restricted flow, however, allows sufficient fluid passage to permit the wheel to turn slowly.

The actual resistance to shimmy is the preload of the accumulator. As mentioned before, fluid under pressure exerts more force on a given area than fluid that is not under pressure. Therefore, when the nose-wheel strikes an obstruction, the wheel, through linkage, throws a sudden added load onto the fluid in the shimmy damper, because the fluid is restricted and cannot flow freely through the orifice.

An example of this resistance may be best understood by comparing the orifice in the damper to the nozzle of a fire hose. When the nozzle is partially open, the hose is hard, and when the nozzle is restricted a bit more, the resistance increases and the hose becomes still harder. The hardness of the hose indicates that pressure is being built up as the resistance increases.

In a similar manner, the faster and harder the nose wheel tries to oscillate, the more difficult it becomes for it to do so.

Piston Shimmy Damper Maintenance

It is imperative that preload be checked on all shimmy dampers each day, and this inspection should not merely concern the preload, but should include as well a thorough check of linkages for tightness, proper lubrication, wear, and damage. Very often, shimmy will be caused by improper installation of a tire, a loose wheel bearing nut, or even nose-heaviness of the plane. Include a check for leakage in the daily inspection.

To fill and bleed the piston shimmy damper, release air pressure by loosening the valve body one or two turns, and hold with a wrench. When air has escaped, disconnect the pistons and pull the rods all the way out. Remove the gage, then move the pistons slowly back and forth to remove air.

When all air has been eliminated, pull the piston all the way out and fill the unit with specified hydraulic fluid through the fitting from which the gage was removed. Then replace the gage and reinstall the valve body in the accumulator with a new crush washer and a new high-pressure valve core. Preload the accumulator with specified air pressure.

Attempts to move the piston rods should meet with immediate resistance if the bleeding has been successful. If such test proves satisfactory, connect the piston rods to the linkage, then clean and lubricate the mechanical linkage.

VANE TYPE SHIMMY DAMPER

Another type of shimmy damper is the vane type, designed for installation either externally or within the inner cylinder of the nose-wheel shock strut.

External types of vane shimmy dampers are fastened to the outer cylinders of the shock strut with mounting lugs which are a part of the shimmy-damper housing. The rotating vane shaft is connected to the inner cylinder of the shock strut by a series of linkages—a lever, a link, an actuating lever integral with an oscillatory collar or slip ring on the shock cylinder, torque scissors, and the wheel fork.

This series of linkages transmit any angular movement of the wheel through this linkage to the damper shaft. Usually, the total allowable oscillation in this type of mounting is 120 degrees, which permits an adequate turning radius.

The wheel spindle, supported by a horizontal arm integral with the strut piston, is mounted fore or aft of the shock strut. The damper is mounted on this horizontal arm. Integral with the wheel spindle is the actuating arm. The spindle actuating lever swivels through 360 degrees and acts as a crank which oscillates the damper lever and shaft through approximately 120 degrees.

An oscillatory pair of vanes connected to a shaft

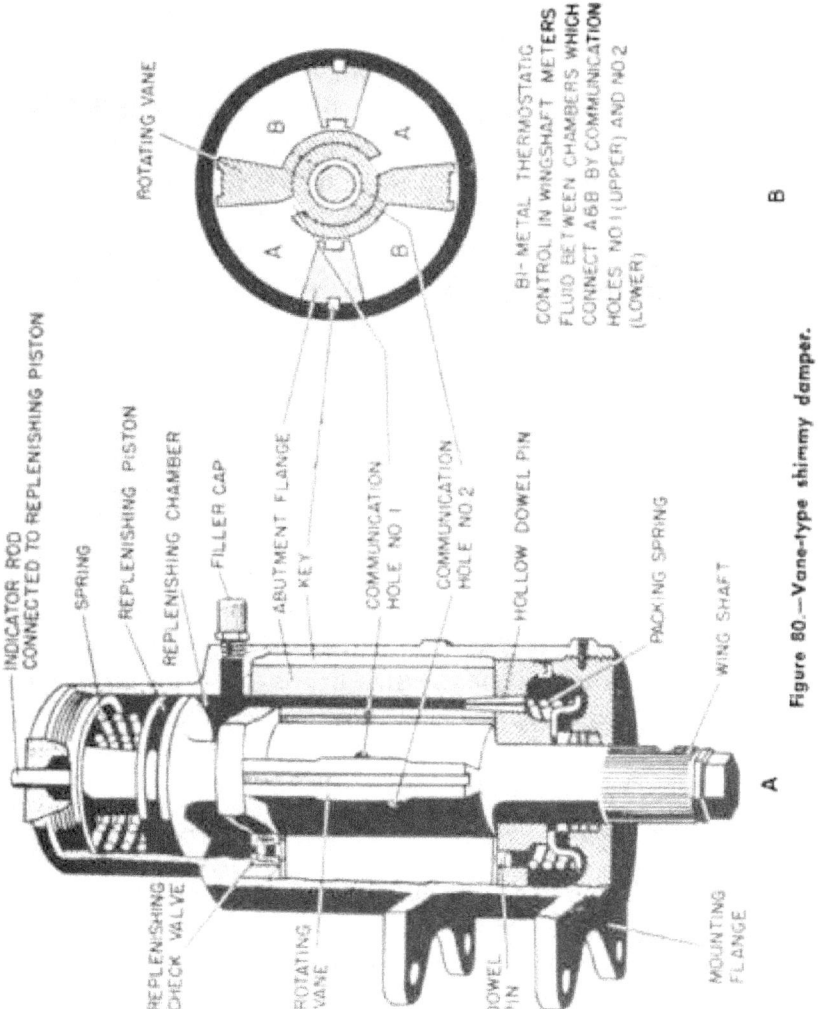

ROTATING VANE

INDICATOR ROD
CONNECTED TO REPLENISHING PISTON

SPRING

REPLENISHING PISTON

REPLENISHING CHAMBER

FILLER CAP

ABUTMENT FLANGE

KEY

COMMUNICATION HOLE NO 1

COMMUNICATION HOLE NO 2

HOLLOW DOWEL PIN

PACKING SPRING

WING SHAFT

REPLENISHING CHECK VALVE

ROTATING VANE

DOWEL PIN

MOUNTING FLANGE

BI-METAL THERMOSTATIC CONTROL IN WINGSHAFT METERS FLUID BETWEEN CHAMBERS WHICH CONNECT A&B BY COMMUNICATION HOLES NO 1 (UPPER) AND NO 2 (LOWER)

A

B

Figure 80.—Vane-type shimmy damper.

known as the wingshaft, and a stationary pair of vanes known as the abutment flange, keyed to the reservoir, constitute the working chamber of the vane-type shimmy damper. This combination forms two pairs of diametrically opposite pressure chambers. Displacement of fluid and consequent resistance is obtained by the movable wingshaft turning toward the stationary abutment flange.

Fluid is passed through a valve orifice, through port holes in the wingshaft, from the decreasingly formed chambers to the adjacent increasingly formed chambers, as the position of the wingshaft changes in relation to the abutment flange. The control valve is externally adjustable, the adjusting element being protected by a removable cap or cap nut. The valve orifice is of variable rectangular shape, and is formed by the relative position of two slots, one in the valve plug, and one in the valve spool.

Section B of figure 80 is a sketch showing the wingshaft with respect to the stationary vanes or abutments. One port through the hub connects two pressure chambers and another port connects the remaining two pressure chambers. The valve connects the two ports. Rotation of the wingshaft in a counter-clockwise direction will cause the fluid in the pressure chambers to be forced through the ports and valves into the opposite pressure chambers. Since all pressure (fluid) must pass through the valve, the resistance is controlled by the size of this valve opening. The adjusting screw on the serrated end of the shaft is used to adjust the orifice to increase or decrease flow.

Study the markings carefully before attempting an adjustment. Note the factory-setting mark and the letters 0 and S stamped on either side of this mark on the shaft. An indicator arrow is also stamped on the adjusting screw, pointing directly toward the factory-setting mark. Turning the indicator arrow counter-clockwise, or toward the O, will close the orifice and restrict the flow, thereby increasing resistance. Turning the indicator arrow toward the S will open the orifice, thereby lessening resistance.

It may be necessary to readjust this setting when moving from warm to cold climates, and vice versa.

The orifice has been designed for thermostatic control. A bimetal element, sensitive to temperature changes, has been installed to take care of viscosity due to operating temperature changes of the fluid.

One of the characteristics of a hydraulic system is that the resistance is proportional approximately to the square of the velocity. Because of this, the shimmy damper has very little resistance to slow motions such as steering, but has great resistance to violent motions such as shimmy. Resistance is offered in both directions of rotation.

All shimmy dampers have a fluid replenishing chamber located in the small end of the damper opposite the protruding wingshaft, and connected to the working chamber by means of one-way spring-loaded valves. Suction in the working chambers opens the valves and draws fluid from the replenishing chamber. Because of the necessity of keeping these valves immersed in fluid, and of keeping the replenishing chamber full, even when the damper is inverted, a spring-loaded piston is used.

Vane-Type Shimmy Damper Testing

To test these units for proper operation, disconnect the torque scissors of the internal type at the center, and the torque arm at the link on the external type. Using the linkage as a lever, endeavor to rotate the vanes. If the damper is properly working, the unit should offer immediate resistance, and should turn smoothly. If immediate resistance is not met, the unit probably requires bleeding.

Shimmy damper bleeding is accomplished by removing the damper and placing it upon a solid mounting (not a vise) with the Zerk fitting straight up. Crack the Zerk fitting and hold until the pressure is released. Then remove the fitting and fill the replenishing chamber to overflowing with specified hydraulic fluid. Rotate the vanes to remove air, and repeat the procedure until no air is present. Replace the Zerk fitting and fill a Zerk gun with fluid. Pump this fluid into the unit until the indicator rod extends as specified on the metal tag attached to each damper.

QUIZ
1. Of what should the daily inspection of shimmy dampers consist?
2. How does the snubber valve operate?
3- How are external types of vane shimmy dampers fastened to the outer cylinders?
4. What is the wingshaft?
5. What characteristic of a hydraulic system causes the shimmy damper to have very little resistance to show motions such as steering, but great resistance to violent motions such as shimmy?

CHAPTER 17

HYDRAULIC SYSTEMS TYPES OF SYSTEMS

The Aviation Structural Mechanic is not directly concerned with the designing of aircraft hydraulic systems. However, in order to understand clearly the reasons for the incorporation of various systems in Naval aircraft, it is necessary that the AM consider several factors pertinent to system design.

Take the matter of weight, for example. Weight is the foremost factor in aircraft design and operation, and it is of the utmost importance that weight be held to the barest minimum. Each superfluous pound added to the hydraulic system, or to any other part of the airplane,

is a pound which must be dropped from the maximum useful load.

Therefore, having established the necessity for considering system design factors, let's delve into the whys and wherefores of hydraulic systems to determine the reason for the varied designs.

An efficient hydraulic system demands that an adequate source of hydraulic fluid be supplied under a pressure greater than that which is likely to be required in the operation of the airplane. The pressure is regulated and the fluid is metered into the .various hydraulic units as it is needed.

An Aeronautical Specifications require that aircraft hydraulic systems be of the type in which the system pressure will return to a back pressure of not over 300 p. s. i. at the pump at a temperature of 0° F. (minus 18° C.) by either manual or automatic operation. These conditions are desired during all operating conditions, but are mandatory during "in flight" operations. ("In flight" operations do not include landing or take-off).

Hydraulic systems are classified as open center systems and closed center systems. There are two kinds of closed center systems: pressurized systems and non-pressurized systems.

Pressurized hydraulic systems are so designed that when the operation of a unit requiring low pressure has been completed, the system will automatically return to operating pressure.

Non-pressurized systems are designed so that when the operation of a unit is completed during "in flight" condition, the entire system is returned to the maximum allowable back pressure.

Two open-center systems and two closed-center systems are included in the non-pressurized category, as follows:

Open-center — variable delivery pump control. This system is under pressure only when one or more services are operated.

Open-center — constant delivery pump pressure relief valve control. This system is under pressure only when one or more services are operated. Closed center — variable delivery pump control. Only during "in flight" operations, when none of the services are being used, is this system depres-surized. Automatic devices are used for depressur-izing naval aircraft, while air force planes use either an automatic or manual device. Closed center — constant delivery pump-pressure regulator CONTROL. This system is depressur-ized only during "in flight" operations when none of the services are being used. Two closed center systems are included in the pressurized system category, one controlled by a variable delivery pump, the other by a constant delivery pump and pressure regulator.

Hydraulic systems in naval aircraft fall into one of three classes, depending upon the cut-out pressure at the main pressure controlling device—3,000 p. s. i., 1,500 p. s. i., and 1,000 p. s. i.

Open Center System

As with all other systems, the open center hydraulic system uses a reservoir to house a reserve supply of fluid; an engine-driven pump to provide fluid flow; a filter to clean the fluid; a main system relief valve to protect the system from excessive pressures, and a system pressure gauge to indicate the degree of pressure. At this point, all similarity between this and other hydraulic systems ceases.

The open center system contains open center or automatic neutral selector valves instead of the conventional piston or poppet-type selector valves. When these valves are in neutral, the output of the power pump circulates through them to the reservoir.

These selector valves are connected in series rather than in parallel. That is, all open center system selector

valves are installed in the same pressure line, and fluid passes through each valve before returning to the reservoir. When one selector valve is put into the working position, the others are inoperative. In an open center system, the selector valve is the unit which relieves the pump of load when the mechanism reaches the end of its stroke.

Closed Center System

The closed center or pressure regulator hydraulic system has two additional units which the open center system does not incorporate. A pressure regulator, or un-loader valve, is used to regulate system pressure and to take the load off the engine-driven pump when no units are in operation. The cut-out pressure of the regulator is set at 100 to 350 p. s. i. lower. The system is under pressure at all times while the engine is running, but when system pressure reaches the cut-out pressure setting of the regulator, the regulator "cuts out" and diverts pump flow back to the reservoir under free flow, thus unloading the pump.

An accumulator is used in conjunction with the regulator, its purpose being to aid the regulator; to absorb shocks and surges; and to aid the pump during peak demands by storing fluid under pressure. It is obvious that if an accumulator were not used, the regulator would continuously cycle because the internal leakage of a few drops of fluid would lower system pressure several hundred pounds per square inch. With an accumulator, the leakage is replaced by fluid stored under pressure in an accumulator.

Other units in the pressure regulator system are similar to those employed in the open center system. These include the reservoir, engine-driven pump, filter, system pressure gage, and main system relief valve.

Power brakes may be used on a center system because the system is always under pressure when the pump (or pumps) are operating. If the engine is not running, the hand pump must be used to provide brake pressure.

Various provisions may be made for emergency hand pump operation of unit systems, but the use of hand pump selector valves and shut-off valves usually are restored to for such action.

The hand pump selector valve is used to isolate each of the unit systems from the others when so desired. Thus the entire hand pump output may be directed to any one unit system, such as the landing gear, wing flaps, bomb-bay doors, etc. This is necessary in case of system failure so that undamaged parts of the system may be operated.

A manifold block, with built-in check valves and thermal relief valves, is used in conjunction with the hand pump selector valve. The check valves prevent hand pump flow from going to any unit system except that selected, but allow engine-pump flow to pass to all unit systems. Thermal relief valves relieve thermal expansion when fluid is trapped in unit systems.

Shut-off valves are used in another type of closed center system to isolate one or more sections of the system. The hydraulically-operated units needed to effect a safe landing are in one section, called the primary system, and can be isolated from the secondary system by closing the shut-off valve, thus directing the hand pump output exclusively to the primary system. The primary system included landing gear, wing flaps, and power brakes. The remainder of the system units are in the secondary system.

EMERGENCY HYDRAULIC SYSTEMS

In case of hydraulic system failure, a reserve method of operating essential flight units must be provided. By flight units is meant those elements necessary for airplane landing, such as landing gear and wing flaps.

Current aircraft models employ several means of emergency landing gear extension for use under various conditions. In case of partial hydraulic failure, for example,

DO g
w I
« c
So
« a
O

Automatic Pressure Regulator—Accumulator System (closed center).
a. Regulator cut-out pressure; Upper limit of operating pressure for all units (regulator cut-in pressure)
b.
c. System relief valve setting, cracking pressure

0)
S §
CO ^
I -
1 > I >
egg
| §
CU P.

Closed or Open Center System (Variable Volume Pump)

a. Maximum pressure for all units, pump cut-out

b. Upper limit of full flow (system) pressure

c. System relief valve setting, cracking pressure

41
a>
to o
SB
C
eg
cp
3
C
CP
3
to
CO
a> u a
CP
C 3
CP
C 3
C CP
•*->
CO >> -A.
o
CP
3
CO
m
CP
p.
3
co to <p
P4
CP
X.
CO o
I!
cd S
CP
u
3 ^
o to .£
p ?> "5
CU CU w
-T3
<V
G
■

-《-》
c
8
S w
cc ><
60
.J P <
Q
P
w
09
o o
°.
CO
o z <
o
o
o o o
2
OS
p in
2f ∎
-*->
a
to
X u •^-《
CO O
3
-*-> o co
C CJ a>
0)
cy
0)
O
C
co
CN

Hose burst pressure to
be in accordance with ap-
plicable detail specification. if lower

e o

s
CO
cy
a> Z
to o
x «
cy m
2 °
cy
co o
c u o
cy
ft «
57 ^ to
*4
I«8
o
o o o ic
o o o
.CO
co
U3 y V ft
o
w o (V o
ft
c '5
0) CO 0)
8 £
rH Q>
CO CO CD
ft
o o
Wo
CO CO
w
:) £ 2 X*
Efl O
_ . cy cd cy
2U S& g.«g B
CO
CO
2 p,
©
© o o o o w
o o
CO

—- cd
C 3
cy
cy cy
g «
hi co
CO CJ
1 s.
o o ©^
o
CO
o o
CO
r
•8
□ 03
CO o
B I
o o o
o
•4-* CO
B 3 . J to
y - C3 ^ O
*i C S O cy cO a
cu
CO CO

_
o o o
o o
o o
o ic
— CM
*-< CO
o £ 5 ..
o £ o co
co ^ co
ft
CO
cy
w|
CO
Pi c
Q 4)
So <~
Slj 2
O

o
fa
o
to
CO w ft
||
CO
C
CO
co Q>
CO
O
K
*>
CO
-
cd q)
C to co
C cu
M
CO Q.
c
C _ P o
1^
bo ^
c
C
•«
cd ■*->
c o
0J
cy
co
5^
_ 3
« g
el £3
2
to S C ^ cy
s
8 a

the hand pump would furnish the required pressure. Two sources of emergency operation are available in case of complete hydraulic failure—extension by gravity, and the use of some auxiliary force such as compressed air or carbon dioxide.

The gravity-type SYSTEM is used where the design and construction of the aircraft will permit the landing gear to be dropped by reason of its own weight. This is accomplished by release of the up-locks through a cable-linkage arrangement allowing landing-gear weight to carry it down. Positive down-locking is insured on some planes, where landing-gear construction permits, by a spring-loaded actuating cylinder which snaps gear into the down-locking position.

Emergency systems using compressed air or carbon dioxide are basically the same with regard to units and principles of operation. For the sake of simplicity, we shall discuss the operation of a compressed air type system.

The compressed-air system consists of a control handle located in the cockpit, to which a cable arrangement is attached to control unit operation. When the handle is placed in the down position, four distinct actions occur: cable movement causes the air-vent to close and seal the air system; a bypass dump valve is opened to permit fluid to bypass to the return; the tail wheel is unlocked, allowing the spring to snap it down, and the air bottle pressure is opened, releasing air to the shuttle valves which seal off the normal fluid-entry port and allow air to enter the actuating cylinder and extend the landing gear.

Air-bottle pressure is sufficient for one emergency operation only, and must be reserviced after use.

QUIZ

1. AN Specifications require that aircraft hydraulic systems be of a type in which the system pressure will return to a back pressure of not over how many degrees at a temperature of 0° F. by either manual or automatic operation?

2. What are the two classes of hydraulic systems?

3. What are the two kinds of closed-center systems?

4. What type of neutral selector valves does the open center hydraulic system contain?

5. Are the selector valves of the open center hydraulic system connected in series or parallel?

6. What two additional units does the closed center hydraulic system have that the open center system does not have?

7. What two sources of emergency operation are available in case of complete hydraulic failure?

CHAPTER 18

HYDRAULIC SYSTEM INSPECTION AND MAINTENANCE

GENERAL

Constant and thorough checking are required to maintain any hydraulic system. Every unit and part in the system must function perfectly if the system is to operate efficiently. To test the system or its parts, a power-driven pump is required- This pump, in most hydrauliq systems, is engine-driven. To test these systems, without operating the airplane engine, an auxiliary source of pressure is required.

Two standard auxiliary sources of pressure for testing hydraulic systems are currently in use. The gasoline-driven test stand is used wherever electric power is not available. Electrically-driven test stands are commonly used for testing entire systems, sub-systems, or individual components.

If neither of these stands is available, an emergency stand may be constructed.

OPERATIONAL CHECK OF HYDRAULIC SYSTEMS

An operational check of all parts of the system, except landing gear, may be performed with the system pump or either type of test stand. Before beginning a landing gear check, the airplane should be properly supported so as to allow operation of the landing gear without harm to the aircraft. Check the fluid level in the reservoir and accumulator pre-load, and inspect all lines and units for external leaks.

If inconvenient or unsafe to use the system pump, a test stand must be used. Connect the stand pressure and return lines to the corresponding lines in the system. These connections usually are made at self-sealing couplings at or forward of the firewall. Charge the test stand accumulator and apply pressure to the system by opening the correct valves.

Watch pressure on the test stand pressure gage. If the pressure drops, check all lines and units including the test stand for external leaks, trace all leaks to their sources, and eliminate the cause. If a pressure drop exists and no external leak is found, test various sections of the system until the internal leak is located.

Start the test stand pump and, if possible, adjust the output to agree with that of the system power pump. Set the test stand relief valve or pressure regulator higher than the system relief valve, after which the airplane system may be operated in the usual manner.

Operate all mechanisms through at least two cycles. Check the time and pressure required for operation against the time and pressure specified as normal. Note whether or not the time required decreases with successive operations. Stop all mechanisms in an intermediate position, and note whether or not the mechanism creeps. The upper operating pressure of unloading valves may be checked by putting all selector valves in neutral. If the system has a pressure regulator, relief valves may be tested by using the hand pump or by removing them from the system.

Check the position indicators against the position of the mechanism through a complete cycle. Check the charge of air in the accumulator, and if low, check the accumulator for leaks.

If the time required for operation is greater than that specified as normal, test stand pump delivery may be below rated, an internal leak may exist, or air may be present in the system. If the time of operation decreases with each cycle, air is being worked out of the system. "Spongy" action of the hand pump denotes air in the system, while prolonged, excessively easy operation of the pump often indicates an actual leak in the pump.

If the pressure attained with the hand pump fluctuates, there may be air in the system, or the oil supply in the reservoir may be low. If the mechanism operates with a jerky motion, binding or fouling of the linkage or cylinder may be indicated. If the accumulator cannot be charged to the operating pressure because of the pump "cutting out," the pressure regulator is probably set too low.

ELIMINATION OF TROUBLES

Now let's consider several methods of eliminating a few of the troubles usually found upon operational checks. Any serious trouble thus found must be eliminated before restoring the aircraft to active duty.

External Leaks

If the pressure drops or fails to rise, immediately inspect all lines and units for external leaks. If an external leak is found at a connection, and the coupling nut is found to be loose, tighten the coupling nut gently. If gentle tightening fails to stop the leak, disconnect the line and inspect the flare for cracks and embedded grit and the mating surface of the fitting for scratches. Serious overtightening of a flare may stop a leak, but will probably make the flare so thin that it will fail under vibration in flight. Therefore, never overtighten a flare

connection. Never use a "bigger" wrench. If a packinr or gasket is leaking, tighten the packing nuts. Should the leak persist, replace the packing or gasket.

Internal Leaks

If an internal leak is found, localize the trouble by testing sections' of the system. When the section containing the leak is located, test the units in the section. Remove the faulty unit, and, if a replacement unit is not available, repair the leakir.g unit. Test it, then reinstall in the system if the leak has been checked.

Air In System

If air is found in the system, bleed the system by operating all mechanisms. It may be necessary to operate all units through several cycles before all air is eliminated.

Faulty Linkage

If linkage is binding or fouled, repair or replace the faulty portion. Care should be exercised in straightening bent linkage.

Low Air Pressure In Accumulator

If the charge of air in the accumulator is low, first check for an internal leak. If no leak is evident, check for an external leak.

Incorrect Operating Pressure

If the accumulator cannot be charged to operating pressure, first check the fluid level in the reservoir. If this is- not the trouble, check the setting of the pressure regulator and adjust the valve if the setting is incorrect. Regulators which are not adjustable must be replaced.

HYDRAULIC SYSTEM MAINTENANCE

All aircraft hydraulic maintenance begins with inspec-
tion. This procedure involves checking the general condition of all component parts of the system, checking for the presence of leaks, and checking the operation of system units. The

general procedure for daily pre-flight, 30-hour, 60-hour, 90-hour, and 120-hour inspections, or checks, are outlined in the following paragraphs.

Pre-flight inspection is performed daily before the first flight. This is chiefly a visual check and includes inspection of the landing gear struts for proper extension; reservoir for proper fluid level; operation of one hand pump unit system to observe relief pressure of the main system relief valve, and checking the accumulator preload pressure. The brakes are also checked for presence of air in the brake system.

Periodic checks —such as the 30-hour, 60-hour, and 120-hour inspections—are made for the purpose of checking the operation, security, and general condition of all hydraulic parts. Procedure and specific items to be checked may be found in The Erection and Maintenance Manual for each type of aircraft.

FLUSHING A HYDRAULIC SYSTEM

It is possible that, through error, a vegetable-base fluid may get into a hydraulic system fitted with synthetic rubber packings and designed for the use of mineral-base fluid. This mistake will cause considerable damage, the most serious consequence being that vegetable oil will cause swelling of packings, gaskets, and hose lines designed for mineral-base fluid.

Vegetable oil, when mixed with mineral-base fluid, forms a gummy residue. In case the hydraulic system should respond sluggishly, a sample of the oil with which it is filled should be tested for the presence of castor oil and alcohol found in vegetable-base oil.

In any event, the ten following steps should be observed when flushing an aircraft hydraulic system:

1. Fold the wings of the plane and secure them with jurystruts or cables. Then hoist the plane so that

the main landing gear and tail wheel may be readily retracted and extended.

2. Drain the system thoroughly. Open the lines at all cylinders, with the exception of the wing-fold —in which case it is advisable to disconnect the flex hoses at the wing-hinge swivels. If possible, move the pistons in and out by hand.

In some instances, this will require disconnecting the piston-rod ends. In the case of the wing-fold strut, disconnect the piston rod and allow it to remain disconnected until later, as will be explained below. Disconnect the gun chargers and open the line at the wing-hinge level, allowing the fluid to run free. Disconnect the pressure line at the main manifold and allow it to drain. Empty the reservoir completely by disconnecting the suction lines from both the engine-driven pump and the hand pump. In draining the engine-driven pump, disconnect the line at the pump, not at the reservoir.

3. When the system has thoroughly drained, reconnect all lines and fill the reservoir with Stoddard Solvent. Disconnect the return line at the reservoir and lead the line into an overflow can. In the case of a dirty system where the actuating cylinders have sustained no damage, the two working lines may be connected together and the actuating cylinder bypassed.

4. Now, using the hand pump, open each selector valve until all cylinders have moved a minimum of five complete cycles. Maintain the reservoir level by adding flushing fluid.

5. Make certain that the arresting hook is fully extended and retrieved; the wing flaps operated to their full throw; the landing gear fully extended and retracted; and the gun chargers manipulated through complete charging cycles. It is important to remember that the charging system must

be bled during this operation, as is explained in the section below which deals with the filling of a hydraulic system.

6. Operate the wing-fold selector and pull the engine through about 25 revolutions to flush the line from the engine-driven pump to the manifold. Make certain that the wing-fold piston has remained disconnected throughout this procedure, and that the operation of cylinders merely extends or retracts the piston rods.

7. Now drain the system thoroughly, as before.

8. When drainage has been completed, connect all lines except the return line to the reservoir. Fill the reservoir with the proper hydraulic fluid. When replenishing a system after flushing, lead the return line to an overflow can and continue adding clean fluid to the reservoir until approximately 20 gallons of fresh oil have passed through the system. The overflow can is never to be used for refilling.

9. Now connect the return line to the reservoir and attach all piston-rod ends. Operate all units in sequence to bleed out air, taking care not to allow the landing gear and wings to drop before the cylinders have become filled with fluid.

10. Finally, check the system for leaks and proper operation.

FILLING A HYDRAULIC SYSTEM

The first step in filling a hydraulic system which has been emptied completely is to hoist the plane sufficiently to permit full retraction and extension of the landing gear. The reservoir should then be filled with new, clean hydraulic fluid of the type specified on the instruction plate which is attached to or near the reservoir.

If a portable test stand capable of producing airplane system pressure is available, it should be connected to the aircraft hydraulic system. If self-sealing coupling connections are provided in the fuselage of the plane, the test stand pressure and suction lines should be attached to them. Otherwise, the self-sealing couplings in the engine pump pressure and suction lines should be uncoupled and the system pressure and- suction coupling valves attached to the mating valves in the test stand pump pressure and suction lines.

If no test stand is available, the airplane hand pump may be used.

Making certain that the reservoir is full at all times, successively operate all hydraulically actuated parts. For example, retract the landing gear under hydraulic pressure. (Retraction must take place first, because damage will result if the gear is extended with empty cylinders). Then extend the gear and fill the reservoir.

The next step is to fold the wings hydraulically until the cylinders are properly filled with oil, exercising great care to prevent the panels from falling free. This operation will be more safely performed if several assistants support the wing panels. When the cylinders are filled, spread the wings and refill the reservoir, repeating this performance until all hydraulically actuated parts have been manipulated and the entire system is filled with fluid.

When this procedure has been completed (if a portable test stand has been used), the self-sealing couplings in the test system and pressure lines should be uncoupled and returned to their original condition.

Your attention is directed to the fact that the above procedure does not actually bleed the system. Minute quantities of air will remain trapped at various points. This, however, will not impair efficient operation except as pertains to gun chargers, which require absolute bleeding of air.

Gun chargers may be bled by first disconnecting the flex hose at the charger and then turning the control valves to charge and pressing inward. While an assistant operates the hand pump, elevate the end of the flex hose and close the fittings with light finger pressure until the tube is filled with liquid. Raise the charger to the tube and make the connections so that no fluid is lost-After bleeding, the chargers should be operated through several cycles.

When the system has been filled, the oil level in the reservoir must be maintained at the position indicated on the gage-rod. Fluid level is measured with the airplane in three-point attitude and with wings spread, flaps closed, and landing gear extended.

QUIZ

1. What two standard auxiliary sources of pressure for testing hydraulic systems are currently in use?

2. When making an operational check, if the time of operation decreases with each cycle, what is happening?

3.. Why is excessive tightening of a flare to stop a leak not recommended? *

4. How may an internal leak be localized?

5. Why is vegetable-base hydraulic fluid not used in a hydraulic system fitted with synthetic rubber packings?

6. What is the first step in filling an aircraft hydraulic system which has been emptied completely?

7. How is the type of hydraulic fluid for a particular system determined?

8. Which hydraulically operated mechanism on a plane require absolute bleeding of air for efficient operation?

CHAPTER 19

HYDRAULIC TURRETS AND GUN CHARGERS PURPOSE OF HYDRAULIC TURRETS

When military aviation was in its infancy, a Navy gunner could go aloft and maneuver his guns without much loss of efficiency. This was because airplanes had not developed the terrific speeds common in modern aviation. As new and improved types of planes were developed, however, and as increases in speed and maneuverability became perennial innovations, the gunner found it increasingly difficult to track and follow aerial targets. He also realized that he was taking physical beatings from the slipstream.

It thus became evident that some means be provided whereby the gunner could maneuver his armament rapidly and accurately and at the same time be protected from air currents.

Gun turrets were the answer to this problem, and thus the reason for this important phase of our discussion.

TURRET CODE

Quite a number of manufacturers suppy naval aviation with hydraulically actuated turrets, and each type is identified by a code designation which supplies the name of the maker, the amount and type of armament, the source of power, and the modification number.

To illustrate, let's examine the Martin 250-SH-2 turret. The name Martin naturally means that the turret is a product of the Glenn L. Martin Co. The number 250 indicates that the turret is armed with two .50 caliber machine guns. The S informs that the turret is spherical in shape, while the H means that it is hydraulically operated. The numer 2 is merely the modification number.

The Consair 250-CH-4 turret, to consider another example, is manufactured by the Consolidated Aircraft Corporation, has two .50 caliber machine guns, is cylindrical in shape, and is hydraulically-operated. It is the fourth modification of the model.

The Erco 250-TH-2 is a product of the Engineering and Research Corporation, has two .50 caliber guns, is shaped like a tear-drop (T for tear), and is actuated hydraulically.

TURRET NOMENCLATURE

The nomenclature used in describing turret operation is listed below.

Elevation —guns moving up. Depression —guns moving down. Azimuth —guns rotating horizontally. Trunion —axle or pivot point.

Swivel —the unit used to transfer fluid from a stationary to a moving part.

Dead man's switches —electrical switches closed by pressure of the gunner's hand. Should the gunner be injured, the relaxing of his grip will automatically shut off the turret. These switches are mounted on the handles of the control valve. Fire interruptor —automatically stops the guns from firing when the gunner brings the turrets within range of his own aircraft. Limit stops —prevents guns from hitting the aircraft structure.

TURRET CLASSIFICATION

Turrets are mechanically classified according to their shape. First, there is the ball, or spherical, type in which the guns are fixed to the turret and the gunner moves with the armament in both azimuth and elevation training. The second is the cylindrical turret in which the gunner and turret move in azimuth only, and the guns pivot in elevation.

Hydraulically, turrets are classified into those types haying valve control and those controlled by a pump.

In the valve-controlled system, the speed and direction of the turret is controlled by handles operating a hydraulic control valve which governs the turret in azimuth elevation and depression. Two pumps are required for the pump-control system—one for azimuth and one for elevation and depression.

MARTIN 250-SN-3 TURRET

Because of its simplicity of design and operation, let us first consider the Martin 250-SH-3 turret. This unit carries 800 rounds of ammunition, 400 rounds for each of its .50 caliber machine guns. The guns are charged and fired electrically, but a foot-trigger permits emergency charging and firing.

The hydraulic system of this turret is a self-contained unit supplied by a variable-displacement pump driven electrically. This pump, described in detail later in this chapter, operates at maximum volume up to a pressure setting which may be adjusted to any point between 600 and 900 p. s. i. An increase of 100 p. s. i. above the pressure setting results in the volume being reduced from maximum to zero. The pump housing serves as a reservoir, and has a built-in supercharging chamber which renders the system independent of atmospheric pressure.

Since the system must be completely sealed, return flow must equal the pump delivery, thereby designating this a balanced system. The pressure is normally regulated by the compensation valve, but should this valve fail, a relief valve built into the pump will limit the pressure to 1100 p. s. i.

After leaving the pump, the fluid flows to a Clarke Control Valve, which is nothing more than a housing containing two four-way selector valves. One valve controls movement in azimuth, the other in elevation and depression. Both handles are operated by hand grips attached to the valve by a control column.

The dead-man switches are attached to each hand grip. Either or both of these units must be depressed to operate the pumps, and the pump must function to energize the trigger circuit. The trigger switches—one or both of which will fire the guns—are also attached to each of the hand grips.

The elevation drive consists of a gear box, a hydraulic motor, and a 90-degree manual-drive gear box. The hydraulic motor is the seven-piston constant displacement type, and will move the turret 77 degrees in elevation and 33 degrees in depression. In case manual operation is desired, move the elevation clutch lever to the up position and rotate the manual control handle.

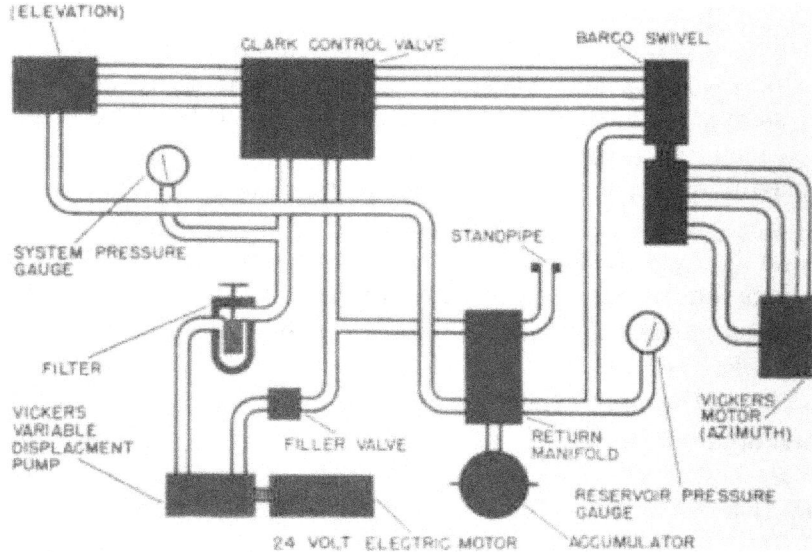

Figure 81 .—Schematic diagram of Martin 250-SH-3 hydraulic turret.

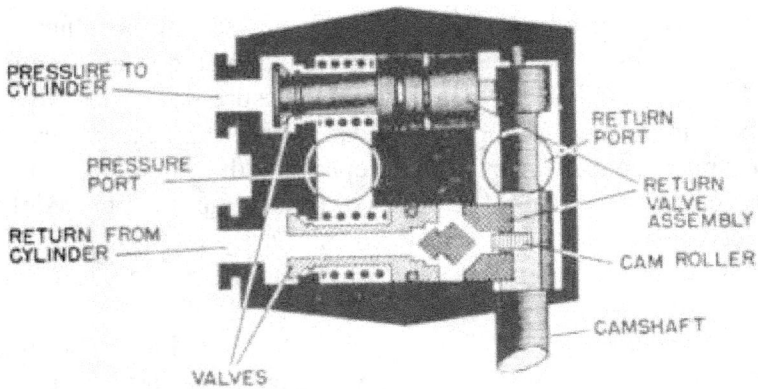

Figure 82.-Clarke Control Valve.

In the DOWN position, the manual control is inoperative, and the turret then is actuated hydraulically.

The azimuth drive is almost identical with the elevation-drive unit. The turret rotates in azimuth 90 degrees to either side of the aircraft centerline. Operation of the turret—manually or hydraulically—is controlled by a clutch lever. In the forward position, the turret is ready for hydraulic operation, while it may be manually actuated in the after position.

In the pressure line between the pump and the control valve is located a line filter which cleans the fluid. The system pressure gage and snubber are located in the same line.

A return manifold is installed in the return line to which a five-inch accumulator with a preload of 25 to 30 p. s. i. is attached. This provides a positive feed pressure to the pump, this pressure being indicated on a low-pressure gage.

A filter valve is installed so that the system may be filled and bled in one operation. This valve must be in the run position when the turret is being operated, and must never be turned to the fill position unless the system is to be filled and bled, and unless the filler valve is properly attached to a pressure tank. If the turret is operated with the filler valve in the fill position, back pressure would build up in the drain lines to the motors and the housing gaskets would be blown out.

A swivel is mounted in the right-hand trunion to transfer fluid in three lines from the ball to the saddle. Two of these lines run from the control valve to the azimuth motor, the third being a drain line from the motor housing to the low-pressure side of the pump.

MARTIN 250-CH-6 TURRET

Now that we have an idea of what to expect in a hydraulic turret, let's consider the Martin 250--CH--6 turret as either an upper deck or a tail unit. Basically, the hydraulic system is the same as that just discussed,

so we will mention only those units not incorporated in the Martin 250-SH-3 turret.

The guns in the 250-CH-6 turret are charged hy-draulically by Bendix gun-charging valves and cylinders, which will be discussed in the latter phase of this chapter.

A contour-follower valve is installed in the elevator working lines of the 250-CH-6 unit. When the guns approach too near the fuselage, this valve is energized by a switch to prevent the guns from striking the fuselage. To prevent this valve from moving the guns too rapidly, a restrictor is installed in the pressure line leading to the valve.

MOTOR PRODUCTS 250-CH-5 AND 6 TURRETS

The Motor Products cylindrical hydraulic turret has two modifications. The 250-CH-5 unit is used as a nose turret, and the 250-CH-6 operates as a tail turret.

The hydraulic system is the same in both modifications. Power is supplied from a small power panel called the ACTUATOR UNIT mounted outside the turret. This panel incorporates a reservoir, an electrically driven gear pump, a check valve, an electro-hydraulic pressure switch, a filter, and a relief valve.

Flow in the actuator unit begins at the reservoir and goes to the pump. From the pump, fluid passes the relief valve, a check valve, into and past the pressure switch, and enters the accumulator. From the accumulator, fluid travels to the swivel and on into the turret where it operates the actuating units and returns through the swivel, through a filter, and back to the reservoir to complete a cycle.

When the pressure reaches 1,100 p. s. i., the pressure switch breaks the electrical circuit and stops the pumps. When the pressure drops to 900 p. s. i., the pressure switch turns on the pump. If the pressure switch fails to cut out, pressure is relieved at 1250 p. s. i. by the relief valve. When the pressure switch stops the pump, the

check valve prevents stored fluid pressure in the accumulator from motorizing the pump.

The accumulator is preloaded to 600 p. s. i., and prevents excessive cycling of the pressure switch, thus creating smoother operation.

After entering the turret through the swivel, fluid passes through a shut-off valve. This valve, which prevents accidental operation of the turret and injury to the gunner because of a charged accumulator, must remain closed while fluid enters and leaves the turret. From the shut-off valve, fluid enters the Clarke Control Valve and is directed to the hydraulic units operating the guns in elevation and the turret in azimuth. Return fluid from the control valve passes through the swivel, through the filter, and into the reservoir.

To operate the guns in elevation and depression, two actuating cylinders, called gun jacks, are provided. The speed and movement of the jacks is controlled by the Clarke Control Valve.

A dump valve is installed in the working lines to the jacks for the purpose of allowing fluid in both lines to flow freely from one to the other, or into the return, when the open position is indicated. This releases pressure and permits the manual controls to be operated without working against pressure.

To move the turret in azimuth, an azimuth actuating CYLINDER, controlled by the Clarke Valve, is provided. A dump valve is installed in the working lines for the same purpose as in the elevation lines.

HYDRAULIC GUN-CHARGING SYSTEM

Since this chapter is the only place in our study of aircraft hydraulics touching on aviation armament, let us consider the subject of the hydraulic gun-charging system before concluding this phase.

Bendix Gun-Charger Control Valve and Cylinder

The purpose of the gun-charger control valve is to pre-

ELEVATION AND GUN JACKS
DEPRESSION

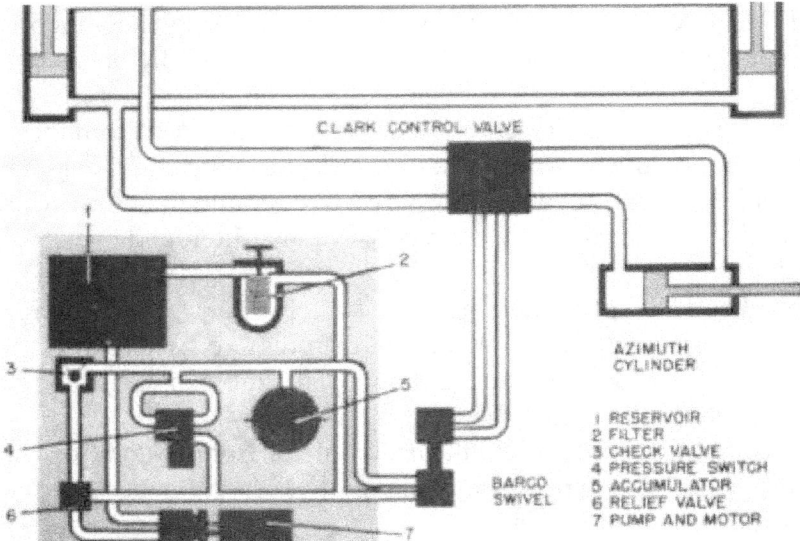

Figuro 83.-S<hematic diagram of Motor Product. 250-CH-5 hydraulic turrot.

pare quickly and remotely a machine gun or cannon for firing by directing pressure to an actuating cylinder which operates the gun breech through its first cycle. The unit also may be used to safety the gun by trapping fluid in the actuating cylinder at the end of the first half of the cycle.

The Bendix gun-charger valve is more widely used at present than any other type. The unit illustrated in figure 84 consists of a three-port housing, two check valves, an operating handle, and a plunger. One check port is located in the inport, the other situated in the cylinder port. A pin lies between each of the check valves and the plunger, in one end of which is a relief valve and a stop. Shoulders on its top extend through holes in the plunger. A small spring-loaded detent plunger is mounted in the housing.

Each 20-mm. cannon has one hydraulic gun-charger

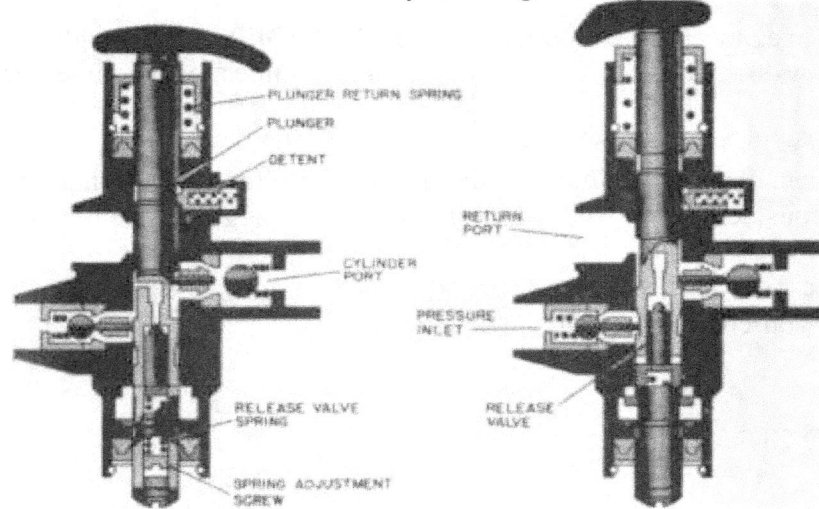

CYLINDER BEING ACTUATED "SAFE" POSITION

Figure 84.—Schematic diagram of a Bendix gun-charger valve.

cylinder mounted so as to form an integral part of the gun. This unit is mounted in the cylindrical portion on the left side of the cannon housing, and is used to charge initially. Another function of this unit is to render the gun safe or inoperative by holding, or retaining, hydraulic pressure in the cylinder. This forces the breech to open position, where it will be held so long as hydraulic pressure exists in the charging line.

The gun-charging cylinder is nothing more or less than a single-acting spring-loaded unit—such as we covered in our discussion of actuating cylinders.

Gun-Charger System Operation

To charge the guns automatically, the control handle is turned to fire position and pushed inward. This action directs fluid flow to the gun-charger cylinder. When the gun bolt is in the fully retracted position, pressure builds up to 750 p. s. i. and the valve automatically kicks out, allowing fluid in the gun-charger cylinder to return to the reservoir. When fluid pressure is released, the spring in the gun-charger cylinder extends the piston, permitting the bolt to close.

If it is desired to keep the bolt retracted, the valve handle is turned to the safe position before it is pushed inward. When in this position, the gun-charger valve holds the pressure locked in the charger cylinder to hold the bolt open indefinitely. To close the bolt, the valve handle is turned to fire, in which position the valve permits pressure in the charger cylinder to escape back to the reservoir.

The gun-charger valve may be used for charging only three .50 caliber guns at one time, the interval being longer to actuate additional charger cylinders. The valve cannot be used in a system having more than 30 p. s. i. back-pressure in the return line. This difficulty is overcome by installation of a separate return line to the reservoir for the valve. The reason for this is that the spring in the cylinder will not push the fluid back out of the cylinder if back-pressure greater than 30 p. s. i. exists. Failure of fluid to flow back sufficiently fast to furnish the bolt with a snap action in the battery position necessitates a mechanical lock to retard the fluid until the piston is clear of the bolt.

QUIZ

1. For what two primary problems were gun turrets the answer?
2. For what does the 250 stand in the Erco 250-TH-2 turret designation?
3. What term is applied to turret guns' horizontal rotation?
4. What is to keep a gunner from firing on his own plane?
5. Name the two mechanical classifications of turrets.
6. Name the two hydraulic classifications of turrets.
7. To what position must the elevation clutch lever of the Martin 250-SN-3 turret be moved to actuate the turret hydraulically?
8. How are the guns in a Martin 250-CH-6 turret charged?
9. To operate the guns of the Motor Products 250-CH-5 and 6 turrets in elevation and depression, two actuating cylinders are provided. What are they called and by what is their speed and movement controlled?
10. What device makes possible remote control of a machine gun or cannon?

CHAPTER 20

AUTOMATIC PILOTS PURPOSE

In the first chapter of our discussion, we hinted that hydraulics flew airplanes, and now the time has come for the consideration of just how this takes place. For hydraulics actually does fly aircraft, and, aided by air and electricity, does a smoother and more efficient job than human hands may ever hope to perform.

We refer, of course, to automatic pilots. The primary purpose of these devices is to stabilize an airplane about its three axes and maintain it on its set course—regardless of whether that course be straight, circular, or helical. The automatic pilot is mostly used for holding the aircraft on a set course in long flights.

Basically, the hydraulic systems of autopilots are very similar in design and, unit for unit, are very similar to the main aircraft hydraulic system. In fact, autopilots on planes equipped with main hydraulic systems utilize the main system as a source of pressure.

S-3 AND J-l AUTOPILOT SYSTEMS

So let's trace the flow of fluid through the Sperry S-3 and the Jack and Heintz J-l autopilot systems and

compare their operation to those systems and units already discussed.

In previous studies of hydraulic systems, we found that the first and foremost unit to be considered was the reservoir, or sump. In planes using a main hydraulic system, the fluid flow is drawn from the system reservoir, and in most cases the fluid supply is furnished by the main system pump.

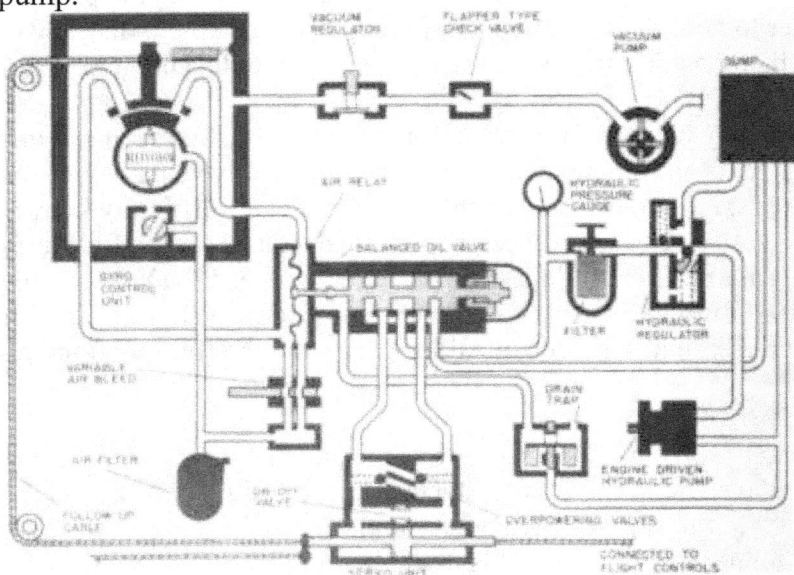

Figure 85.—S—3 and J-1 autopilot systomt.

When filling an autopilot reservoir, carefully check the fluid specifications stamped on a metal tag attached to the unit. In separate autopilot systems, the fluid used is termed Servo Oil, a mineral-based fluid listed as Specification AN-00-366A.

Some types of aircraft have separate autopilot pumps powered either as an auxiliary engine or by an electric motor.

An important unit required in most autopilot systems is a valve to control pressure. Autopilot systems normally operate at pressures ranging between 90 and 200 p. s. i., depending mainly upon the plane in which they are installed. The pressure-regulating valve in the S-3 autopilot is a VlCKERS HYDRO-CONE BALANCED RELIEF VALVE.

A balanced relief valve is used because of its smooth operation, since it is absolutely essential that autopilot systems receive an even and constant flow of fluid when in operation.

If a system is to operate efficiently, clean fluid is essential, and therefore a filter is installed in the main pressure line. These units should be cleaned with the use of solvents.

The pressure gage is followed by an on-off BLEED (shut-off) valve, although this unit is replaced in certain autopilots by a check valve. An on-off valve is also located in the servo assembly which, when open, bypasses fluid from one end of the balanced actuating cylinders to the other to permit operation of the controls when the autopilot is not in operation.

In our studies of hydraulic systems, we learned that the next unit in sequence of operation was the selector valve. In autopilot systems, this valve is controlled by vacuum or by electrical impulses, and is called a balanced oil or transfer valve. The selector is of the balanced piston type and has five ports—a pressure port, a return port, a drain port, and two cylinder ports. The drain port leads to the reservoir or, in some models, to a drain trap and thence to the reservoir. A drain trap is installed only where the sump is too high to permit gravity drainage from the mounting assembly to the oil sump.

The J-1 autopilot incorporates a mechanism called the variable air-bleed valve which controls speed by restricting air flow to a diaphragm in the air relay valve, which will be discussed with the vacuum system later in this chapter.

Balanced Oil Valve Adjustment

The balanced oil valve should be removed from the aircraft for adjustment or conditioning. To perform this operation, remove the control boxes from the S-5 installations to permit access to the air relay which must be disconnected from the spool. Care must be tak?n not to damage the pin hooked into the spool or the shaft on the air relay diaphragm when removing. Removal of the end cap will allow for finger pressure to be applied to the spool spring by pushing the spool nut to a position where the two units may be unhooked without damage to the shaft or pin. It is not necessary to remove control boxes of J-1 installations because of a clip arrangement which locks the shafts together. Remove the spool assembly from the casting with a spanner wrench and check for scoring, scratches, and corrosion, both on the spool and in the sleeve.

When assembling the oil valve, the spool assembly must be balanced. To accomplish this, assemble the unit loosely and hold the spool while turning the nut inward

to to

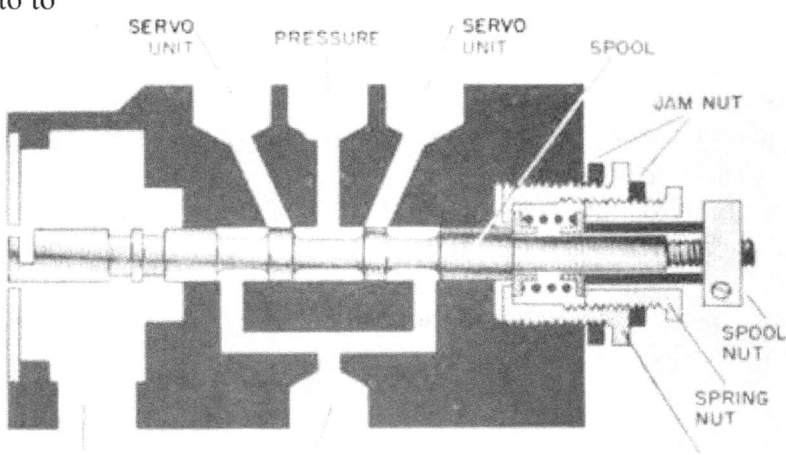

DRAIN RETURN CENTRALIZER
NUT

Figure 86.—Balanced oil valve.

until it barely bottoms on the sleeve. This arrangement must be accurate—neither too tight nor too loose. Without permitting the nut to turn, hold the retainer nut and screw this spring nut into the retainer nut until it also bottoms on the spring seat.

In the final step, hold the unit by the retainer nut and shake to check for end play. If no end play is evident, secure the lock nut without allowing the adjustment to slip, and apply specified system fluid to the spool and landed surfaces before screwing into place.

To check for operation following assembly and installation, turn the on-off valve to the off position and rev up the engine to 1,000 r. p. m. with the speed-control valves closed. Then check for normal operating pressure, which will vary with different types of planes.

Open the speed-control valves to 3-5 settings and turn the autopilot on. The servo shafts should be moved to a hard-over position. Turn outward on the spool assembly until the servo unit and the controls commence to move slowly in one direction. This brings a corresponding action to the follow-up system and pulleys. If, after four or five turns, no servo action is evident, check for the cause of trouble.

At the point where the follow-up begins to creep, work the casting and jam-nut. Screw inward on the retainer nut, and at the point where the follow-up pulley begins to move in the opposite direction, mark the nut. The two points marked indicate where servo action begins. Bisect the angle between these points for location of the adjustment at which no servo action exists. After securing the jam-nut, mark the servo shaft and check for creepage. This position should be held for at least three minutes.

Servo Unit

The next unit in the conventional hydraulic system is the actuating cylinder which, in the autopilot system, is called a servo unit. Three of these units are actuated by three balanced oil valves, and each servo unit operates

the elevator, aileron, and rudder, respectively. Servo units in the S-3 and J-l autopilots are double-acting balanced actuating cylinders with piston rods connected by clevis and shear-action fittings to control cables.

Installed as integral parts of each servo unit are two relief valves called over power valves which are adjusted to relieve at approximately 90 percent of the autopilot system pressure. These valves permit the human pilot to overcome the system manually in case of emergency.

One of the accepted methods of adjusting over-power valves is the bleed method. This operation is identical with the setting of any relief valve, and may be accomplished according to the following outline.

Turn the autopilot system on and adjust the SYSTEM pressure control unit until a gage reading equal to the pressure at which the over-power valves relieve is achieved. Put the controls hard over, and when the piston has reached the end of its travel, turn both adjusting screws all the way inward. Remove the return line from the servo unit and install a short flex line for drainage. Turn the adjusting screw on the extended shaft side outward one-half turn at a time until all leakage shows from the return line into the drainage pan. Kill system pressure, then connect the cylinder port to the system and repeat the operation for adjustment of the opposite side. Readjust the system regulating valve to specified system pressure.

The servo unit also has a bypass valve which the pilot opens by means of a system of cable linkage in order to control the plane without moving fluid through the system. This valve merely bypasses fluid from one end of the actuating cylinder to the other, and is closed when the autopilot is in operation.

Autopilot Vacuum System

Now that we have discussed the salient features of the S-3 and J-l hydraulic pilot systems, let's take a look at

the "brains" of the entire mechanism—the vacuum or air system.

First of all, let us observe the VACUUM PUMP. This unit is usually a vane-type pump, but in some installations a piston pump is employed. Air is drawn from an inlet throughout the system and expelled from the outlet port of the pump. Therefore the logical starting point for our study of this unit is at the inlet port, from whence we can trace the flow of air through the pump.

From the filter, air is drawn into a control box through a "nozzle" and thence directed into cups on the outer circumference of the gyro rotor to turn the rotor at a speed of about 1,500 r. p. m. The principle applied here is much the same as that used in the operation of water turbines. The spinning action of the rotor creates centrifugal force which causes the rotor to hold its position relative to its axis much as a spinning top tends to remain upright while in motion. This tendency to maintain its position is called rigidity in space.

One rotor in the B.C.U. control box governs the longitudinal attitude (bank) and the lateral attitude (climb). Another control box houses the D.C.U. rotor governing the vertical (turn or direction) attitude. Each rotor is mounted in a cased gimbal ring. As the vacuum pump draws air from the box, atmospheric air passing through the filter enters the rotor housing through a jet, causing the rotor to spin at high speed.

The gimbal ring housing is so connected by linkage to the control box that when the airplane swerves off course and rotates about the axis of the gimbal ring, air selector valves called pick-offs increase the vacuum and air pressure on opposite sides of the air relay diaphragm. - The S-3 is called a displacement-type automatic pilot because the speed at which it is brought back on course depends upon the degree off course it has been thrown. For this reason, the S-3 autopilot is used in larger planes. The speed at which it will return the plane to normal flight attitude is controlled by a nozzle and nozzle-

plate in the S-3, and by a shroud and port in the J—1.

Variable Air Bleed

As previously explained, the variable air bleed is a device designed to control the sensitivity of the J-l autopilot. This unit is mounted directly beneath the air relay. When the control knob is turned, the openings will open and close to direct more or less air and vacuum to opposite sides of the diaphragm, thus rendering signals from the gyro rotor stronger or weaker, depending upon flight conditions.

Our discussions have shown that responses of actuating cylinders to flow are instantaneous. We also know that when steering a large ship, the pilot must "go to meet her," to prevent the vessel from veering off in the opposite direction. This also is true of aircraft.

Each of these conditions is provided against in the autopilot by an arrangement of cable linkages called the follow-up system. We have seen that by means of the control-box pick-offs (signal system), the autopilot collects weak (air or pneumatic) signals and transmits, through a series of air relays and balanced oil valves, strong hydraulic signals to the servo units.

All this action is a means of placing a signal into the autopilot system, but the original signal must be cancelled out in order that the signals from the control box be picked up. For the sake of clarity, let's suppose that an actuating cylinder is centered. If the selector valve handle is moved so as to put the piston over to the end of its travel, the handle must again be moved to return the piston to normal or neutral position. But to stop this piston in neutral, the selector handle must be moved once again. That is the action of the follow-up system.

The follow-up is a cable arrangement extending from one end of the servo unit through a series of pulleys to the mounting unit. A clutch device attached to the pick-off assembly will, through servo action, cancel out the plane's original displacement.

These cables may be installed to pull either clockwise or counterclockwise so as to provide the most direct installation with the required relation between control surface movements.

Spiral springs in the pulley drums must pull, or unwind, against the cable. This action is similar to pulling out a spring-loaded flexible steel rule. Drum springs are reversible, and pulleys may be used for either direction of rotation by starting the cable at the inner or outer side.

S-4 AUTOMATIC PILOT

The S-4 autopilot operates, in principle, exactly as those previously discussed. There are, however, certain units—such as the transfer valves and the servo units— which appear different, but which function similarly.

So let's trace the flow through the system and determine exactly what these differences consist of as applied to the units concerned.

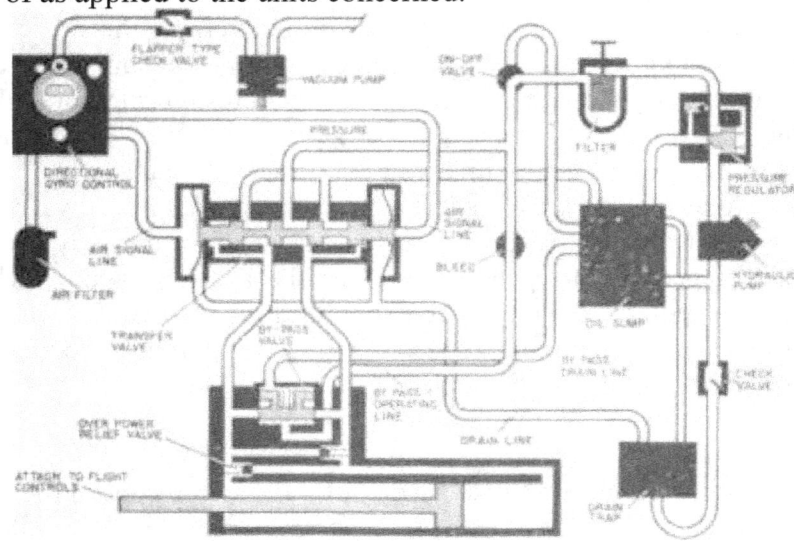

StXVO UNIT

Figure 87.—Schematic drawing of an S-4 automatic pilot.

First, inevitably, is the sump, followed by an engine pump, a pressure regulator, and the main on-off valve, as shown in figure 87.

Transfer Valve

The transfer valve assembly is really a composition of three separate valve assemblies, each complete in itself and each identical in construction and operation. The valve assembly incorporates a balanced piston selector valve operated by two diaphragms.

Servo Unit

The servo unit in the S-4 autopilot is a push-pull type, but is nonetheless an ordinary double-acting unbalanced actuating cylinder. This valve also contains over-power valves which function and are adjusted exactly as these in the S-3 and J-l units.

The S-4 autopilot is called a rate gyro because the speed at which the plane is thrown off course determines the speed at which it will return. For this reason, the S-4 is used on smaller and faster aircraft.

The rate of return is controlled by a bellows arrangement attached to an auxiliary rotor in D.C.U. and the main rotor in B.C.U. When the plane is thrown off course, rigidity in space will hold the gyro in a position compressing or extending the bellows, thus sending a stronger air signal (bypassing the pick-ups) to respective sides of the diaphragm.

G-l AUTOMATIC PILOTS

The hydraulic system of the G-l autopilot, manufactured by General Electric, employs the same units as those systems just discussed. The signal and follow-up systems are, however, electrically controlled.

Our discussion of the G-l autopilot begins with the transfer valve—the only hydraulic unit which cannot be interchanged with corresponding units in the S-4 hydraulic system. The purpose of the transfer valve is to transform electrical signals from the gyro rotor-controlled electrical pick-offs into hydraulic signals which will operate the servo.

The transfer valve assembly contains three double-acting, solenoid-operated, balanced-piston selector valves which control fluid flow to and from the three servo units. When current in one of the two opposing solenoids exceeds current in the other, the associated valve moves to direct fluid to and from the servo unit in the quantity desired.

The valve is arranged for mounting in either a horizontal or a vertical flat surface, with the axis of the valves in a horizontal position.

It stands to reason that the gyro cannot be permitted to send out signals which will jerk the plane back on course. Such signals must be controlled, as in all other autopilot systems, from the servo units by a follow-up system.

Therefore, an electric follow-up is mounted on each of the three servos to control the amount of servo stroke for a given signal and return the plane to a normal flight attitude. The follow-up is little more than a transformer which, when the control switch is moved, will induce more or less current as desired. The control switch is attached by mechanical linkage to the servo piston rod. When the piston rod moves, the transformer induces more or less current to the pick-off to which it is wired. The pick-off will in turn send proportional electrical signals to the solenoids of the transfer valve concerned until the plane is back on course.

In other words, the control returns to normal, or neutral, at the same speed with which the plane is returning to course.

5. a. $A = F/P$ b. $P = F/A$ c $F = P \times A$

6. Pressure exerted on any part of a confined liquid is transmitted to all parts of the liquid regardless of the shape of the container.

7. No.

8. None.

CHAPTER 3

FLUIDS, SEALS, AND TUBING

1. By checking its instruction plate.

2. No.

3. Mineral-base is red; vegetable-base is blue.

4. Castor oil.

5. By the specification number printed or stamped on the can.

6. Because of damage to packings.

7. Toward the pressure or flow.

8. Soft aluminum.

9. Rigid tubing and flexible hose.

10. By color bands-blue-yellow-blue bands.

11. Eighty-two percent.

12. Three times the outside diameter of the tubing.

13. Make certain that the replacement tubing is of the same type of metal, the same size, and that it is formed to the same shape as the original line.

14. This procedure may cut the flare.

15. Nine times the outside diameter.

16. Five to eight percent of its total length.

17. One-piece and two-piece connectors.

18. They make possible the disconnecting of lines without loss of fluid or entrance of air.

19. At the firewall and at other locations where frequent disconnections are made.

20. By hand only.

CHAPTER 4

RESERVOIRS AND FILTERS

1. To supply the operating needs of the hydraulic systems and replenish fluid lost through leakage.

2. At the highest point, usually at the top of the forward firewall or in one of the engine nacelles.

3. It maintains a reserve supply of fluid for emergency hand-pump action, and prevents foreign matter from reaching the engine pump.

4. They prevent excessive vortexing and surging of fluids, and thus prevent air from entering the pump supply lines.

5. Pressurized reservoirs.

6. The open-type.

7. Screening.

8. Cellulose.

9. To filter dust and other contaminating substances from the air.

10. From outside to inside.

CHAPTER 5 HYDRAULIC PUMPS

1. They are too slow and their flow is pulsating and intermittent.

2. The presence of air in the pump or system.

3. At each engine change.

4. Constant-displacement and variable-displacement.

5. The fluid output of the constant-displacement pumps is constant for any rotational speed of the pump, but the fluid output of the variable-displacement power pump is varied to meet the demands of the system.

6. A shear-pin or shear-section.

7. By removing as many 0.001-inch shims from beneath the cover as are required to adjust clearances.

8. Do not drop them on to their seats.

9. In overhaul manuals.

10. They create higher pressure.

11. Disassemble and check clearances; check for proper fit of bushings.

12. The seven-piston type.

CHAPTER 6

PRESSURE REGULATORS

1. Regulators.

2. The automatic type and the semi-automatic type.

3. Leakage.

4. Pressure differential.

5. One complete operation of the regulator from cut-in to cutout.

6. To prevent the development of excessive pressure if the pressure regulator fails to function.

7. The internal drain and the external drain.

8. 2,700 and 3,000 p. s. i.

9. Ten drops per minute. 10. Manually controlled.

CHAPTER 7 SELECTOR VALVES

1. To control the direction of operation of a mechanism.

2. By directing fluid under pressure to the desired end of the actuating cylinder while at the same time directing fluid from the opposite end of the actuating cylinder to the reservoir.

3. Two-way selector valves and four-way selector valves.

4. Pressure port, return port, and two cylinder ports.

5. In-line and coaxial.

6. A piston which is a hollow spool-shaped plunger within the valve housing.

7. It returns to neutral automatically when actuating cylinders have reached the end of their strokes.

8. Relieve thermal expansion from cylinder lines, insure full operation of units actuated, and start piston on its way back to normal from a working position.

9. 150 p. s. i.

10. By screwing the unit in completely and then backing off one-eighth turn.

11. Its ability to prevent more than three degrees of flap-creepage in either direction.

12. 1,000 p. s. i.

13. Because the balls act as poppets in controlling fluid flow through the unit.

14. The return poppet opens two degrees before the opposite pressure poppet opens.

15. 1,500 p. s. i.

16. The number two or general position.

17. 2,500 p. s. i.

18. To allow all poppets to be seated.

19. No external leakage and an internal leakage of no more than 10 drops per hour.

20. Do not drop them on the seats.

CHAPTER 8 ACTUATING CYLINDERS

1. To convert fluid energy into mechanical energy.

2. A cylinder, one or more pistons and piston rods, and the necessary seals.

3. For lubrication and for examination of bushings and mounting.

4. The 10 percent rule—plus 10 percent above, minus 10 percent wide.

5. With the lips facing the flow and pressure.

6. Dirt.

7. It might result in improper, erratic operation of the system.

8. The number of turns required for such a removal should be counted and marked.

9. They are spring loaded and have but one fluid port.

10. Pin-pulling actuating cylinders.

11. A check valve designed to restrict flow in one direction while allowing free flow in the opposite direction.

12. Hose tends to straighten under pressure and may shear off in the fittings or loosen the nuts if not installed straight.

13. Five to eight percent.

14. Nine times the outside diameter of the hose.

15. Scoring, bends, and leaking.

CHAPTER 9
ACCUMULATORS

1. To store hydraulic fluid under pressure.

2. Maintains pressure in the pressure manifold, supplies a limited amount of fluid under

pressure to actuating units when the power pump fails to operate, dampens pressure surges, absorbs fluid shocks, and prevents too frequent cut-in and cut-out of a pressure regulator.

3. The air load must always be released.

4. The diaphragm type, the bladder type, and the piston type.

5. By brushing with soapy water.

6. Loss of air pressure, low air pressure, or high air pressure.

7. By replacing the valve core, the gasket or both; or by replacing the diaphragm.

8. Check for leaks and damaged diaphragm, charge to correct preload pressure and check air valve for leakage, bleed to correct valve.

9. Sluggish response of accumulator. 10. By replacing the damaged diaphragm.

CHAPTER 10 LOCKS AND SEQUENCE VALVES

1. Unlock the mechanical lock and then place the selector valve in the unlock position.

2. A unit with a piston shaft through each cylinder cap for the simultaneous operation of two locking pins.

3. Snap-action type.

4. To cause one hydraulic action to follow another in definite order of sequence.

5. The unbalanced type.

6. To allow free flow of fluid when the wing is fully spread.

CHAPTER 11 RELIEF VALVES

•

1. Pressure relief valves.

2. Of a two- or four-port housing, a ball or cone held on its seat by a strong spring, and an adjusting screw.

3. The ball when held on its seat by the strong spring.

4. Main system relief valve and thermal relief valve.

5. Balanced and unbalanced types.

6. To relieve above pressure system pressure.

7. 1,750 p. s. i.

8. A small metering hole in the piston.

9. By turning an adjusting screw.

10. 2,000 p. s. L

11. To prevent damage to cylinders formerly resulting when engine exhaust gases heated the liquid.

12. The higher the compression of a spring becomes, the higher the pressure required to overcome it.

CHAPTER 12 AUXILIARY UNITS

1. Dump valves.

2. Orifice check valves.

3. The manifold block.

4. To limit the rate of flow in both directions in a line.

5. Flow equalizer.

6. The piston type and the gear type.

7. 1,400 p. s. i.

8. To aid the pilot in moving plane controls.

9. An emergency bypass valve.

10. In p. s. i. above atmospheric pressure (14.7 p. s. i.).

11. A pressure-gage snubber.

12. A pressure switch.

13. 260 p. s. i.

14. A dashpot assembly.

15. Hydraulic fuses.

16. Quantity measuring type and return flow type.

17. Eliminates need for excessive mechanical linkage and saves weight.

18. Automatically by pressure switches or thermostats or manually by toggle switches located in the cockpit.

19. A winding around a moveable iron core.

20. The Kenyon Manifold Block.

CHAPTER 13

SHOCK STRUTS

1. Absorb and dissipate the landing shock on the compression stroke of the strut, absorb and dissipate the jolts of taxiing, absorb recoil shocks which occur on the extension stroke of the shock strut on take off, and act as structural members to support.

2. Metering pin.

3. To maintain correct alinement of the wheels.

4. Hard main-gear struts and soft main-gear struts.

5. Hard struts.

6. In the compressed position.

7. Communication holes directly beneath the piston head.

8. Fluid being metered through the flapper valve.

9. The scissors arrangement.

10. The rate at which fluid is displaced from the piston chamber.

11. For dissipation of the energy.

12. Depress the valve core and allow it to snap back.

13. Every 30 hours.

14. Every 120 hours.

15. To prevent binding as the strut extends.

CHAPTER 14 BRAKE ACTUATING SYSTEMS

1. Mechanically, hydraulically, or pneumatically.

2. Admission of system pressure to brakes because of failure of seals or metering mechanisms would allow planes without nose wheels to nose over.

3. Supplies pressure for the brakes.

4. The passage of fluid through the compensating ports.

5. A small ball check in the filler cap.

6. By engaging a ratchet which locks the piston in the ON position.

7. Difference in area.

8. To reduce pressure, obtain greater fluid displacement, faster return, and smoother operation.

9. To release and regulate pressure to the brakes.

10. Pressure chamber, brake chamber, and return chamber.

11. ON, OFF, and BALANCED.

12. A check valve and a piston containing a floating pin.

13. Full operating pressure.

14. By "spongy" action of foot pedals.

15. Using the master cylinder as a pump, by forcing fluid through the system with compressed air, and by operating foot pedals.

16. The OFF adjustment and the HIGH-PRESSURE adjustment.

17. Full operating pressure.

18. By varying the stroke of the push rod.

CHAPTER 15

HYDRAULIC BRAKES

1. Shoe-clearance and bleeding.

2. Two-shoe servo and single-shoe servo.

3. The servo action is effective in single-servo brakes for one direction of wheel rotation only.

4. Because the brake lines are single-static there is no circulation in them,

5. Brake drums not round, brake drums scored, or oil-soaked lining.

6. One.

7. 0.008-inch to 0.010-inch.

8. The brake frame, the expander tube, and brake blocks.

9. From 200 p. s. i. to 300 p. s. i.

10. Taxi plane and apply brakes several times to dry the linings.

11. 0.003-inch clearance per disk.

12. Fixed-disc type and floating-disc type.

CHAPTER 16 SHIMMY DAMPERS

1. Checking the preload; checking linkages for tightness, proper lubrication, wear, and damage; and checking for leakage.

2. It allows free fluid flow when returning to neutral and restricts flow when the wheel tends to swing out of line.

3. With mounting lugs which are part of the shimmy damper housing.

4. An oscillatory pair of vanes connected to a shaft.

5. The resistance is proportional approximately to the square of the velocity.

CHAPTER 17 HYDRAULIC SYSTEMS

L 300 p. s. i.

2. Open center and closed center.

3. Pressurized and non-pressui ized systems.

4. Open center or automatic neutral selector valves.

5. Series.

6. A pressure regulator and an accumulator.

7. Extension by gravity and use of some auxiliary force such as compressed air.

CHAPTER 18

HYDRAULIC SYSTEM INSPECTION AND MAINTENANCE

1. The gasoline-driven test stand and the electrically driven test stand.

2. Air is being worked out of the system.

3. It may thin the flare sufficiently to cause it to fail under vibration in flight.

4. By testing sections of the system.

5. It will cause swelling of packings, gaskets and hose lines designed for mineral-base fluid.

6. Hoist the plane sufficiently to permit full retraction and extension of the landing gear.

7. It is specified on the instruction plate which is attached to or near the reservoir.

8. Gun chargers.

CHAPTER 19

HYDRAULIC TURRETS AND GUN CHARGERS

1. Maneuverability of armament and protection from air currents.

2. Two .50 caliber machine guns.

3. Azimuth.

4. The fire interrupter.

5. Spherical and cylindrical.

6. Valve-controlled systems and pump-controlled systems,

7. The DOWN position.

8. Hydraulically by Bendix gun-charging valves and cylinders.

9. Gun Jacks controlled by the Clark Control Valve. 10. Gun-charger control valve.

CHAPTER 20 AUTOMATIC PILOTS

1. Servo Oil-Specification AN-00-366A.

2. Between 90 and 200 p. s. i.

3. A Vickers hydro-cone balanced relief valve.

4. By vacuum or by electrical impulses.

5. It controls speed by restricting air flow to a diaphragm in the air relay valve.

6. Rigidity in space.

7. Because the speed at which it is brought back on course depends upon the degree off course it has been thrown.

8. The push-pull type.

9. Because it is a rate gyro—that is, the speed at which the plane is thrown off course determines the speed at which it will return.

10. Electrically.

APPENDIX II

QUALIFICATIONS FOR ADVANCEMENT IN RATING AVIATION STRUCTURAL MECHANICS (AM) RATING CODE NO. 810

General Service Rating

Aviation Structural Mechanics maintain and repair aircraft surfaces, structures, and hydraulic systems. Aline structural parts, such as wings, elevators, ailerons, rudders, and fuselage structures. Prepare, paint, or dope aircraft surfaces. Repair rudder, plastic, fabric, and wooden structures used in aircraft construction.

Emergency Service Ratings

Naval Job Classifications

Naval Job Classifications—Continued

Qualifications for Advancement in Rating

Qualifications for Advancement in Rating—Continued

Qualifications for advancement in rating

Applicable rates

.103 BLUEPRINTS

Read simple blueprints and drawings..

Read and work from blueprints and drawings

Make working sketches for structural or hydraulic repair

.104 CONSTRUCTION

Remove, repair, service, install, and aline as appropriate aircraft structures, control rigging, and fittings, including wings, control surfaces, tabs, landing gear, control cables, and fuselage structures .

.105 CLEANING OF AIRCRAFT

Clean aircraft surfaces, structures, and enclosures, using proper materials and procedures. Use steam for cleaning aircraft

.106 METAL WORKING

Identify common aircraft metals, tubing, and fuel, oil, and hydraulic lines. Fabricate aircraft sheet metal parts, metal fittings, and tubing by cutting, flaring, bending, threading, and assembling, t'se riveting tools and riveting machines. Use safety and bending wire where appropriate. Make repairs to metal structures, including stressed skin and frames

Install and maintain hydraulic lines, including the replacement of packing and seals and the repair or installation of flexible hose..

Qualifications for Advancement in Rating—Continued

Qualifications for advancement in rating

Braze, anneal, forge, and otherwise perform metal heat-treating operations encountered in aircraft structural maintenance

Use sandblasting and plating apparatus for preparing metal surfaces, if activity to which assigned is so equipped

.107 WBLDINO

Set up oxyacetylene welding apparatus and perform simple welding and cutting operations on carbon steel

Braze and silver-solder applicable metals. Weld ferrous and aluminum alloys—

Perform simple arc-welding operations on steel plates and tubes... Note.—See welding test instructions under .400.

.108 HYDRAULIC SYSTEMS

Trace through aircraft landing gear, bomb bay, automatic pilot, brake, and other hydraulic systems; repair and service individual parts and linkages as required. Make periodic checks and inspections to facilitate preventive maintenance. Vent, bleed, drain, flush, and refill hydraulic systems. Remove, service, repair, and install hydraulic unite and accessories

Set up, operate, and maintain test benches for hydraulic units and accessories

Applicable rates

AM

810

2, 1, C

2, 1, C

AMS

811

3, 2, 1, C

2, I, C l.C

2, 1,C 1,C

2, 1, C

2, 1, C

3, 2, 1, C

2, 1, C 1,C

AMI.

812

3, 2, 1, (2, 1. (

Qualifications for Advancement in Rating—Continued

Qualifications for Advancement in Rating—Continued

Qualifications for advancement in rating

Applicable rates

AM

810

AMS

811

AMH

812

.113 SUPERVISION

Supervise and train personnel engaged in aircraft structural or
hydraulic repair

Organize and administer:

Metal repair shop

Hydraulics repair shop

xxx .200 EXAMINATION SUBJECTS

.201 TOOLS AND MEASURING INSTRUMENTS

Types, nomenclature, and uses of hand and power-driven tools used in the structural maintenance of aircraft, including those employed in metal working, woodworking, rubber, fabrics, and plastics repairs, painting, and the rigging of cable. Types, nomenclature, and uses of various measuring instruments employed in structural maintenance of aircraft

.202 CONSTRUCTION

Types of aircraft construction. Maintenance procedures for removing, installing, rigging, and alining fuselage, structures, wings, tail surfaces, landing gears, tabs, control cables, cowlings, inspection plates, and fairings. Basic principles of the theory of flight and of weight and balance

1,C

C

c

1.C

c

i,C

3, 2, 1, C

3, 2, 1, C

3, 2, 1, C

3, 2, 1, C

3, 2, 1, C

3, 2, 1, C

Qualifications for Advancement in Rating—Continued

Processes for fabricating and joining metals. Processes and purposes of heat treating, including surface treatment for aluminum and magnesium alloys and corrosion resistant steels

.204 WELDING

Types and characteristics of welding apparatus and of welds. Welding processes, including use of material, technique, and

safety precautions

Note. —See welding test instructions under .400.

.205 HYDRAULICS

Systems of aircraft that are generally hydraulically controlled. Principles of hydraulics for transmission of power. General repair, service, and maintenance problems common to hydraulic systems, including removal, testing, installation, and inspection of various units of such systems as the landing gear, brake, flap,

3, 2, 1, C

3, 2, 1, C

3, 2, 1, C

3, 2, 1, C

Qualifications for Advancement in Rating—Continued

Qualifications for advancement in rating

.205 hydraulics —continued bomb bay, automatic pilot, and booster control. Principles of instruments used in aircraft hydraulic systems. Lubricants and liquids used

.206 WOOD, RUBBER, PLASTICS, AND FABRICS

General properties of wood, rubber, fabrics, and various plastics and their uses in aircraft construction. Processes of repair, inspection, and testing of aircraft structures and fittings made from these materials

Properties and methods of preparing and applying glues and rubber. Vulcanizing processes

.207 PAINTING AND CLEANING

Types, characteristics, and properties of paints, dope, varnishes, lacquers, pigments, driers, enamels, and thinners; and effects on color and properties caused by mixing. Types of hand and air brushes used and the methods of preparing and applying paint to aircraft surfaces. Methods of cleaning and caring for painting equipment. Types, characteristics, and uses of cleaning materials for cleaning aircraft surfaces and enclosures. Materials and methods used to mask surfaces and stencil insignia or numbers on aircraft

Applicable rates

3, 2, 1, C I 3, 2, 1, C

Qualifications for Advancement in Rating—Continued

Qualifications for advancement in rating

Applicable rates

MATHEMATICS

Basic mathematical principles as that apply to aircraft structural maintenance, as follows:

Elementary principles of equations, powers, roots, and proportions. Work an elementary problem in weight

and balance

Basic principles of triangles, squares, parallelograms, circles, and functions of right angles for computing metal shapes and forms

.209 SAFETY PRECAUTIONS

Local and general safety precautions pertaining to shop and line maintenance of aircraft structures, including those to be observed when painting, doping, welding, using power-driven tools, fueling, and otherwise servicing or handling aircraft

.210 SUPPLIES

Basic principles of Navy supply system, including procurement, stowage, custody, issue, and inventory

.211 RECORDS AND REPORTS

Common forms in use and the procedures for their preparation. Records kept and reports made for administering an aircraft

Qualifications for Advancement in Rating—Continued

Applicable rates

xxx.300 NORMAL PATH OF ADVANCEMENT TO WARRANT GRADE

Aviation structural mechanics advance to Warrant CARPENTER 7711 (Aviation Structural Technician) and assist Engineering Officers in repair and maintenance of aircraft structures.

xxx.400 INSTRUCTIONS FOR TESTING AND QUALIFYING WELDERS

Note. —Qualified welders (metal-arc and gas) are divided into three classes: welders, third class; welders, second class; and welders, first class.

.401 QUALIFICATIONS FOB WELDERS, THIRD CLASS

Pass the following qualifications test in accordance with the requirements of the General Specifications for Inspection of Material — Appendix VII — Welding, Part E:

Section E-l: Test No. 1 in vertical and overhead position, using

approved electrodes. Section E-2: Test No. 1 in flat position only on steel, bronze, and

cast iron, using applicable welding rods. Section E-5: Tests Nos. 1 and 2. Pass an examination on these subjects:

Welding symbols, types of welds, nomenclature, and definitions as set forth in sections A-l and A-2 of the General Speciflca-

Qualifications for Advancement in Rating—Continued

Won* for Inspection of Material — Appendix VII — Welding, Part A.

Uijes of cooper, brass, aluminum, Iron, steel, and various alloys

aboard naval vessels, preheat and postheat treatment of metals encountered in welding. Various types of metal-arc welding sets.

Current and voltage necessary for various sizes and types at

electrodes used in metal-arc welding. Proper flames and technique to be used in gas welding and cutting

of various materials, together with proper tip sizes that should

be used.

Safety precautions to be observed with regard to welding, cutting, and to handling of gases used.

.402 TESTS AND QUALIFICATIONS FOB WELDERS, SECOND CLASS

Must have served at least 1 year as welders, third class. Pass the following qualification tests in accordance with the requirements of the General Specifications for Inspection of Material — Appendix VII—Welding, Part E:

Section E-l: Test No. 4 using carbon molybdenum pipe and electrodes; Test No. 1 in flat position only, on nickel-copper, corrosion-resisting steel, and aluminum, using applicable electrodes.

Section E-2: Test No. 3 using steel tubing and welding rods; Test No. 1 in flat position on aluminum, using applicable welding rod.

Qualify to take charge of welding activities aboard ship and lay out work for men on a Job.

.403 TESfS AND QUALIFICATIONS FOB WELDERS, FIRST CLASS
Must have served at least 1 year as welders, second class.
Take charge of a welding shop aboard a tender or repair ship, lay out,
and properly supervise the work. Instruct and qualify candidates for welders, third class and second
class.
.404 QUALIFICATION AND REQUALIFICATION
The period of qualification of welders shall be for 18 months. Quail* flcation or requalification tests will be conducted aboard repair vessels or aboard any vessel having the necessary equipment